Face-to-face Nature Conservation

Academic training often leaves students poorly prepared for what they actually encounter in the field, where bridging the distance to the general public in matters related to wildlife and the status of our planet can pose a real challenge. The increasing disconnect between nature and an often skeptical or misguided public means that many people have a poor understanding of how to behave in the great outdoors and take a bewildered or even confrontational stance against those dedicated to nature conservation and species protection. Managing nature has simply become increasingly synonymous with managing people.

This primer equips practitioners – teachers, students, researchers, citizen scientists, park rangers and guides, zoo staffs, and volunteers – for what they can expect to hear while going about their business. It provides a set of concise answers to over 200 questions or assertions arranged in 20 headings covering major ecosystems and their inhabitants, natural history institutions, and various human perspectives and activities. You'll never again be caught off guard, stumped, or flustered.

A dash of humor and a twinkle in the eye maximizes the chances that those on the other side of the proverbial fence will shake your hand and urge you to keep up the good work. No team should gear up for work without a copy of this book.

"This book will, I feel, help lift its readers from their doldrums and give them some answers and also a smile. It is lucid, witty, well-informed and opinionated in a good way. Ideal for use by undergraduate students, lecturers can also use it to provoke discussion and debate. In the years that I taught Environmental Sciences at undergraduate level, I would have appreciated such a text!"

Mark P. Simmonds, OBE, *Director of Science for OceanCare and Visiting Research Fellow at the University of Bristol, UK*

"It is a major challenge for everyone working in nature conservation and environmental protection projects who are confronted with the general public to provide good answers to questions and to counter sometimes misguided or even provocative statements. This book is an enormous help in this regard and offers a rich source of possible answers. The topics accurately reflect the daily interactions we have as scientists on the front lines with skeptical members of the public."

Patricia Holm, *Emeritus Professor of Ecology, Dept. of Environmental Sciences, University of Basel, Switzerland*

"There is a need for a novel-format book like this one, especially given the challenging times we live in with short attention spans, digital distractions, explosive reactions and temperaments, and escalating conflicts. I admire the novelty of the book and appreciate the intent as well as implementation. The succinct primer is a good jumping board for coming up with custom retorts using humor to respond to challenging questions and diffusing potentially conflict-inducing situations."

Kadambari Devarajan, *ecologist, writer, and National Geographic Explorer*

"This book will fill a void in understanding and communication between various sectors and interest groups involved in work related to animals and the environment."

Michael W. Fox, *veterinarian, medical scientist, and ethologist, syndicated "Animal Doctor" newspaper columnist and author of* One Health: Veterinary, Ethical & Environmental Perspectives, *USA*

"This light-hearted and 'bite-sized' issue-by-issue approach is useful for getting people connected with today's scientific fascinations, but also the pressing issues with which we have to grapple but are doing so extremely sub-optimally today. It has the potential to engage curiosity and motivate with science, but also humor."

Mark Everard, *Associate Professor, University of the West of England, UK*

Face-to-face Nature Conservation

Provocative Questions and Useful Retorts

Michael Stachowitsch

CRC Press
Taylor & Francis Group
Boca Raton London New York

CRC Press is an imprint of the
Taylor & Francis Group, an **informa** business

Designed cover image: Author's own

First edition published 2027
by CRC Press
2385 NW Executive Center Drive, Suite 320, Boca Raton FL 33431

and by CRC Press
4 Park Square, Milton Park, Abingdon, Oxon, OX14 4RN

CRC Press is an imprint of Taylor & Francis Group, LLC

ISBN: 978-1-041-28050-7 (hbk)
ISBN: 978-1-041-28048-4 (pbk)
ISBN: 978-1-003-75747-4 (ebk)

DOI: 10.1201/9781003757474

Typeset in Goudy
by Apex CoVantage, LLC

Contents

About the Author

Michael Stachowitsch is Senior Lecturer at the University of Vienna (Department of Functional and Evolutionary Ecology/Department of Paleontology). He is the author of more than 60 scientific publications (research focus: benthic marine ecology) and two other books, "The Invertebrates: An Illustrated Glossary" and "The Beachcomber's Guide to Marine Debris." In addition to currently lecturing on courses, including Introduction to Marine Pollution and Management of Endangered Species, Michael has engaged in long-term activities managing endangered species, specifically sea turtles. He is President of the Society for Science and Research, Vienna Aqua Terra Zoo (Haus des Meeres), and Austrian representative at the International Whaling Commission (Scientific Committee and Alternate Commissioner).

Preface

I worked for several decades in the summer with my students on sea turtle nesting beaches in Turkey. We were all entirely convinced that our work was valuable, necessary, right, and even righteous. And that stance was backed up by a solid academic background encompassing biology, ecology, nature conservation, and species protection. All of us were embedded in the classical "ivory tower" bubble. Boy, were we surprised when our internally unquestioned drive and enthusiasm were met with skeptical and even hostile resistance from certain segments of the local population. ("Hey, this is our beach!") More benevolent but equally disconcerting were the comments by some of the people closer to us. ("Why don't you do something more reasonable and gainful over summer?") Children also often stumped us with their innocent, out-of-the-box questions. I also encountered a similar lack of understanding in my work with coral reefs and cetaceans. ("What have the sea turtles /cetaceans /coral reefs ever done for us?") ("Why don't you worry about nature in our own country?")

These types of questions point to a clear disconnect between the "experts" and the general public in matters related to wildlife and the status of our planet. This is both a physical disconnect because most folks are far removed from and may never have experienced a truly pristine environment. It's also psychological because today's mindset and hype are all about our modern (digital) lifestyles. Our well-being and all our futures, however, rely on a natural world that exhibits at least some semblance of functioning and fulfils our most basic needs – even if we sometimes take the services that nature provides for granted. This calls for trying to bridge the gap between science and the general public and for convincingly arguing that nature conservation should be very high on our list of priorities. A bit of humor and a tad less finger-waving seem like a welcome and promising approach.

This book compiles over 200 questions and assertions that have been posed to my colleagues and me or that crop up daily in the media. Many are more "in-your-face" than "face-to-face." Importantly, and in defense of our challengers, virtually all such standpoints and interjections – no matter how brazen or cantankerous – have in fact been raised and dealt with in the scientific literature and other trustworthy sources. The people who approach you may feel threatened, suffer from a measure of hopelessness, be misguided or misinformed, or be not fully aware of the overall picture. How will you react and respond? This book is hopefully a step forward in ensuring a better understanding and helping to guarantee everyone's well-being on this wonderful blue planet. I am certain that readers will have a handful of curious entries of their own. Feel free to forward them (along with your own answers and maybe a reference). Any corrections, improvements, or more apt references to my entries would also be appreciated: michaelstachowitsch@chello.at or stachom5@univie.ac.at.

Acknowledgments

My thanks go to the hundreds of students (and the many teaching staff members) who accompanied me to faraway sea turtle nesting beaches and coral reefs and experienced first-hand the often lamentably compromised natural habitats, along with the interesting opinions of tourists and local residents. They and participants in my other classes prompted this book. Thanks also to my many colleagues at the International Whaling Commission meetings who helped me to understand and meld the oft-opposing positions on cetaceans and whaling. The photographs are mostly mine, with additional contributions by Manuel Marinelli (Figure 7), Judith Hannak (13), Andrea Nouak (18), Werner Huber (28), and my student archives (13, 19, 31). Figures 3, 4, 9, 20, and 30 are Shutterstock images. Christian Hochmeister played a key role in designing the book cover. Last but not least, I am grateful to my editor, Alice Oven, for enthusiastically taking up and promoting this project and to the reviewers she chose for their positive responses.

Introduction

> "Ignorance lies not in the things you don't know, but in the things you know that ain't so."
>
> *(Will Rogers)*

This book addresses those on the "front lines" of nature conservation or species protection efforts – those who are likely to personally interact with an often skeptical or poorly informed public. Whether teacher, student, researcher, citizen scientist, park ranger, zoo or museum staff, fieldworker, or volunteer, you will be fielding and responding to the wealth of their questions, claims, assertions, and criticisms. This requires being able to call up biological, (bio-)ethical, political, philosophical, or even religious perspectives.

The natural world is under unprecedented onslaught. Human impacts are increasingly changing the face of our planet, with some experts calling for designating a new era in Earth's history ("Anthropocene"). Conservationists, however, are often confronted with people who are unaware, misinformed, or in denial. Why? On the one hand, it's a question of immediate priorities and interests: many people are rightly fully occupied with making ends meet on a daily basis and have little time and energy to devote to the bigger picture. On the other hand, most folks often fail to fully recognize the severity of the problem because the environmental deterioration and our perception of it are markedly out of synch. This is keenly described by the shifting baseline syndrome, which broadly states that we typically have little understanding of what the natural state is (or, better said, was) long ago. Accordingly, from a more recent, generational perspective, each of us uses what we (or our parents and teachers) experienced as our reference point (baseline). That, however, is often only a blip in the successive historical decline. We therefore tend to underestimate the cumulative damage and to misjudge the urgency of taking action. This is aggravated by the ever-larger disconnect between the general public and natural habitats along with the wildlife they contain. In fact, very few people have probably had the opportunity to experience a truly pristine environment.

These deficiencies create a major challenge for nature conservation and species protection efforts. Two sets of people are affected – and addressed by this book. The first is the experts dedicated to helping save a specific habitat or certain species. They need to understand how their focus fits into the framework of the changes outlined earlier and what the public perceptions are. What should the desired outcome of their work be? What population level or what habitat state and extent? This background understanding is key to tackling the required tasks. These range from scrounging for permissions and funding, often grueling field work, publishing results and conclusions, to interacting with (and trying to convince) decision- and policy-makers, not to mention the general public (the second set of people affected).

The common adage is that the management of endangered species and habitats has increasingly become synonymous with managing people. And the latter, whether they be tourists, fishers and farmers, hunters, local stakeholders, or politicians, are often skeptical to downright hostile toward conservationists, or as some might put it (if they are friendly), "tree huggers." The reasons vary from a perceived infringement on personal freedom, to fear of a potentially constrained lifestyle and reduced personal comfort, to an imagined personal financial burden. Even those who applaud the beauty and efforts highlighted in TV nature documentaries often bristle when their own home turf is involved – basically the not-in-my-backyard pollution-related stance transposed to the nature conservation stage.

Everyone who has devoted time to conservation efforts has unexpectedly encountered or been unpleasantly confronted by such opposition. The folks "on the other side of the fence" all have some reason for their recalcitrance and are typically more than happy to sound out their arguments. One problem that we from the conservation-minded community have is that we are well-trained in our relatively narrow field of expertise and are fully convinced of the legitimacy of our approach, of its (bio-)ethical and philosophical foundations, and of the required actions. Self-questioning and justification among our peers are rarely necessary. Accordingly, many of us are entirely unprepared when faced with statements that come from "far left field" (or should I say "far right field"), from outside our conceptual frameworks, or with off-the-wall remarks fueled by tangential issues. When confronted with such combative stances, even hardened workers are often so taken aback, so unexpectedly pushed outside their comfort zones, that they tend to be at a loss for words. The same holds true for many who are first getting into the field, for example the volunteers that many conservation projects rely on: they are typically equally stumped and unprepared. In any context, knee-jerk reactions and stammering when flustered or in a state of slight panic tend to lead to sub-optimal results. This is not the best way to ensure that everyone parts ways with smiles on their faces.

How to best respond then? With a little preparedness, everyone can go home happy or at least unscathed. Many challengers make claims that are self-serving, misguided, one-sided, narrow-minded, or directly outrageous and even hostile, often based on misinformation or alternate facts. At best, their stance is often fatalistic ("the trend is irreversible, there is nothing I can do personally"). Or the question is being posed by a child, whose thought processes and concerns can lead to very imaginative queries indeed. In every case, the decision on how to respond will depend on a spontaneous, on-the-spot evaluation. The level of your engagement, the detail of the response, your attitude, and comportment should be tailored to your discussion partner(s). This calls for gauging motivations and assessing levels of agitation. We are all very good at this – it's the classical split-second "first impression." A gesture, a stance, a facial expression, or a word inflection from your vis-à-vis can make all the difference. And, importantly, that person will also be making an instantaneous judgment about you. That judgment might well be suboptimal, especially if you are bedraggled from weeks of fieldwork, lack of sleep, poor laundry service, and the many other deprivations of camp life. Keep that in mind when getting set for more important, planned meetings or negotiations.

This is where this book comes into play. It provides potential answers to some of the more frequently heard statements by those sitting on the other side of the proverbial fence, whether that be an information booth, a chance encounter during fieldwork, at a post-scuba dive beer counter, or a negotiating table with higher-ups or "stakeholders." The question you should ask yourself is, "how can I maximize the chances that, in the end, my vis-a-vis would like – at least in their thoughts – to shake my hand and urge me to keep up the good work." Many of the potential questions or claims you may be confronted with, and listed here, are based on personal experience. You and your team shouldn't gear up for work without this book.

Two types of responses are provided. The first one or two (or three) are quick, amusing, flippant, or sassy statements that may or may not directly answer the initial question or assertion ("snappy answers to stupid questions"). These should be delivered with a disarming twinkle in your eye and will hopefully elicit at least a chuckle, help de-escalate an encounter, or be a quick way out of a potential quagmire and help to end a discussion or confrontation that is not promising a positive outcome. The second is a concise, correct, and scientifically supported explanation. The references provided are a mere "random" starting point for those interested in pursuing the issue further. They are more a reassurance that the respective issue is serious enough – regardless of how arcane it may sound – to have been treated in the scientific literature, often dozens or even hundreds of times, including in research journals, book chapters, or entire books. In the case of more than two authors, I have, for the sake of brevity, included only the first author followed by the standard "et al." ("and others") notation. Journal names are spelled out and links provided where available.

The choice of answer will depend on the immediate circumstances: the friendly or unfriendly nature of the encounter, the number of other people wishing to interact, the time available to you, how thirsty you are, and, I'm not necessarily kidding, whether your opponents are armed or not. In one sentence, your general strategy should be to listen attentively, come across as engaged, knowledgeable, positive, and authentically friendly, with the correct balance between emotionality and only moderate use of statistics. It's more of a "food-for-thought" than an "I'm gonna change your mind here and now" thing. A sequence of two or more answers may be justified, for example, when counterarguments are presented. Whenever faced with being backed into a corner argumentatively, it is often helpful to remember two things. The first is the moniker "if you can't solve the problem, enlarge the context." That can mean presenting some broader and hopefully uncontroversial wisdom that all can agree on (e.g., that we are all in the same boat on this wonderful blue planet and we can hopefully work together to make sure that its bounty remains intact for future generations). The second is to insert a dash of humor because humor is the most disarming of approaches. Finally, be prepared to talk about money. This is something biologists – probably based on their miserable pay scales – are unfortunately relatively unversed in, and which they may consider a dirtier subject than pollution. Of course, all the answers provided are merely templates, and each of you will no doubt be able to lend your answers a personal flavor or twist.

Keep the faith (you know you are on the right track) and keep smiling.

Michael Stachowitsch
University of Vienna

1 General

1. **This species is dangerous to human beings, so why protect it?**
   - What better way to get rid of your mother-in-law!
   - "Does your dog bite?" (Inspector Clouseau in "The Pink Panther Strikes Again")
   - Can you blame them for being aggressive after what we've done to them?
   - The species most dangerous to human beings? *Homo sapiens* (after mosquitoes)!
   - So-called dangerous animals, whether they be poisonous or top predators, all evolved that way over millions of years. The success they enjoy always comes to the detriment of the prey they eat or, from another perspective, of the predator who tries to eat them. Simply put, they are perfect so that we can, with a knowledgeable mien, repeat the platitude "My, aren't they just perfectly adapted to their environment." Could the fact that you feel threatened be because you've encroached on their territory or they've been forced to search for food or a home outside their ever-shrinking habitat? (If it's any consolation, humans do not represent the natural quarry of any predator.) The responsibility for most conflicts seems to rest squarely on our own shoulders. Is failing to protect them, or eradicating them, the solution? Hardly.

   Importantly, "dangerous" animals, like all organisms, play a role in an intact, functioning ecosystem. Intervening here has rarely "improved" the environment. Culling or eradicating predators often causes the populations of their prey to explode. This, in turn, creates new problems that then require major long-term management efforts financially and effort-wise to control. Moreover, many poisonous species, from frogs to snakes and fish, produce toxins that are of great interest to the pharmaceutical industry and may end up in products that can save our lives. Major efforts are being made to test such organisms for ingredients for future medications. And that means jobs, something I think

DOI: 10.1201/9781003757474-1

everyone can support. Finally, two additional perspectives as food for thought. First, almost any animal (and don't forget plants) can hurt you. Second, many of the animals people fear the most are not on the top-20 list of killers. Sharks aren't, but dogs rank number 5 (see also entry Nr. 140)! A better understanding of who fears potentially dangerous wildlife and why can help inform management strategies to reduce human–wildlife conflicts. Clearly, such conflicts are a two-way street.

Draheim et al. (eds.). 2015. *Human-Wildlife Conflict: Complexity in the Marine Environment*. Oxford University Press. 196 pp.

Kaltenborn, B.P. et al. 2006. Living with problem animals – Self-reported fear of potentially dangerous species in the Serengeti Region, Tanzania. *Human Dimensions of Wildlife* 11(6): 397–409. https://doi.org/10.1080/10871200600984323

2. **I'm collecting/killing only one individual: that can't hurt, can it?**
   - What's more important: a living and reproducing individual in its habitat or a dried specimen or parts thereof on your shelf at home (or in your freezer)?
   - What are your criteria for killing that specific individual of the population?
   - The purpose of collecting/killing an organism can range from a personal hobby to a museum collection or university research program. The above sentiment is also a common justification brought forward by hunters (see Hunting & Fishing, entries 112–124). And what child hasn't gotten its first hands-on experience with the natural world by collecting an animal? One of the first questions is whether that organism is to be killed or merely observed and then released. In the case of children, for example, the chances are often slim that wild animals they handle or take home in a jar will survive and be returned to nature.

     From a species protection or nature conservation perspective, certain individuals can play an outsized role in helping ensure the survival of a population. That targeted individual may be:
     - important to maintain the genetic diversity of the population;
     - one of the few reproducing females (or the alpha female, which may be the only one to breed);
     - one of the strongest males;
     - an older individual that plays an important role in the "culture" of the species (e.g., a whale or dolphin that conveys behaviors to younger individuals through social learning); and

– a "matriarch" that is a repository of "knowledge" and experience central to the herd (e.g., African savanna elephant).

In other species, the number of individuals has been reduced to such an extent that each one matters. For the tiny remnant population of Maui's dolphin in New Zealand, for example, the human-caused death of even one dolphin (e.g., by entanglement in fishing gear) would increase the extinction risk for this subspecies. And, of course, taking animals from the wild for personal, educational, or scientific purposes is typically strictly regulated and potentially subject to heavy fines for violations. Finally, you are probably not the only one who wants to collect (kill) just one flower, one snail (shell), butterfly, salamander, lizard, coral, or a member or two of some other wildlife population.

CMS Fact Sheet: Animal Culture and the Conservation of Migratory Species. https://www.cms.int/en/publication/fact-sheet-animal-culture-and-conservation-migratory-species

International Whaling Commission, Scientific Committee Report, 2013, Section114.3.2.1. https://archive.iwc.int/?r=3285

Powell, R. 2007. Collecting animals from nature. *Iguana* 14(1): 47–51.

3. **This species is competing with us for food. Do you really want us to go hungry so that it can be saved?**
   - I think it would probably be more accurate to say that we're stealing its food.
   - If it's not jumping on your table or picnic blanket and stealing your lunch, then it's probably not really competing with you for food.
   - Food security is an important issue for many countries, and competition for food resources with wildlife has occasionally been advanced as part of the problem. The case studies range from animals that damage crops to dolphins or sharks that eat the fish already hooked on longlines (depredation). One example is the claim by some whaling nations that whales need to be culled because they eat large amounts of the same fish that are caught for human consumption. That claim has since been debunked. While there may be some overlap in the food of whales and humans, killing those whales will never solve world hunger. Moreover, the populations of many whales are severely depleted after a century of commercial whaling, so today's whales consume only a fraction of the food they once did. Finally, many whales feed in areas that are not subject to commercial finfish fisheries. In most cases, the responsibility for the "food competition" problem (and for solving it) lies squarely on our shoulders, not with the animals whose territories we have encroached

on or whose numbers or normal food supply we have severely reduced, forcing them to seek other food outside their remnant habitat (see also entry Nr. 108).

Kaschner, K., Pauly, D. 2005. Competition between marine mammals and fisheries: Food for thought. In: Salem, D.J., Rowan, A.N. (eds.), *The State of the Animals III: 2005* (pp. 95–117). Washington, DC: Humane Society Press.

4. **The natural environment is being destroyed, but you can't stop progress or turn back the clock, and there are other options to the problem. And I certainly can't do anything as a single person.**
   - Define progress.
   - There will be no future progress if our environment is destroyed.
   - "If you think you are too small to make a difference, try sleeping in a closed room with a mosquito." (probably an African proverb popularized, among others, by Tenzin Gyatso, 14th Dalai Lama).
   - Are you one of those persons who think colonizing Mars is an option?
   - Intact ecosystems are the basis of all life on our planet, including human survival. "We won't have a society if we destroy the environment" (Margaret Mead). We must do everything we can to maintain their basic functions (ecosystem function) and ensure that they help fulfil our basic needs (ecosystem services). While in many cases it is no longer possible to return to a state before major human impacts and exploitation (historical baseline), we can and must at least try to stop or slow down the clock and agree on a level of impact or status that will help avoid major calamity. Who is destined to achieve this? On the one hand, it requires global efforts such as the climate change accords anchored in the Paris Agreement adopted at the UN Climate Change Conference in 2015. On the other hand, Elinor Ostrom won a Nobel Prize in Economic Sciences for demonstrating that, under the right circumstances, small, local communities are able to jointly and sustainably manage their shared natural resources. And, of course, each one of us *can* make a difference. To quote the anthropologist Margaret Mead again: "Never doubt that a small group of thoughtful committed individuals can change the world. In fact, it's the only thing that ever has."

Unger, N. 2012. *Hetch Hetchy Redux: An Effort to Turn Back the Environmental Clock. History News Network.* Internet distribution only. http://historynewsnetwork.org/article/148998

Walton, D. 2009. *Bison Turn Back the Clock on a Patch of Prairie.* Globe & Mail [Toronto, Canada], 19 May 2009, p. A6. Gale Academic OneFile, link.gale.com/apps/doc/A200308033/AONE?u=anon~8de65fd4&sid=googleScholar&xid=a1dcd500

5. **You conservation people are really hopelessly naive idealists and out of touch with everyday life and the real world.**
   - Idealists and optimists live longer and better.
   - No life is possible without the *real* world, namely functioning nature.
   - "It's life, Jim, but not as we know it." (Mr. Spock, original Star Trek TV series)
   - Those well versed in the affairs of man are often sorely ignorant of the workings of nature. (Mike Stahovich)
   - Well, that certainly depends on how you define the real world. But OK, explain what you mean by the real world and show me how it can manage without nature and natural resources. Then we'll take a look at it together and see if it really can successfully maintain itself without ecosystem functions and services. I think you'll see that those of us who are dedicated to protecting nature may be equally or better anchored in the real world than those who feel more at home in the corporate world of industry, real estate, retail, insurance, and finance, including cryptocurrencies – a rather different and often artificial environment. Those business realities, while admittedly also valid and very pragmatic, are nonetheless ultimately dependent on a functioning natural world. Human well-being and survival – not to mention that of the world's fauna and flora – are fundamentally challenged in an impoverished or severely impaired environment. So, instead of environmentalism being branded as a coercive ideology and environmentalists as naive fundamentalists, perhaps it would help to consider the overall framework or circumstances that ultimately enable human prosperity and, indeed, our very existence. Environmentalists may well be the ones who will enable you and future generations to continue "going about with their business."

   Cross, F.B. 2002. The Naive Environmentalist, 53 Case W. Rsrv. L. Rev. 477–495. https://scholarlycommons.law.case.edu/caselrev/vol53/iss2/13

   https://www.durhamprobonoblog.co.uk/post/eco-fascism-the-dark-side-to-environmentalism

   Under the motto "know thy enemy" (Sun Tzu, ca. 5th century BC). https://www.aei.org/carpe-diem/why-naive-environmentalism-is-like-religious-fundamentalism/

6. **Many more animals are killed, and more habitat is destroyed by natural disturbances, so the human influences are really comparatively minimal.**
   - The newly discussed era in Earth's history, namely the "Anthropocene" – it's totally our show!
   - . . . grant to us the serenity of mind to accept that which cannot be changed, courage to change that which can be changed (namely our

behavior), and wisdom to know the one from the other . . . (modified excerpt from the Serenity Prayer by Reinhold Niebuhr).

- Natural changes and disasters versus human-caused or -abetted ones – the distinction should be clear. Natural changes generally take place slowly and over long periods. This gives organisms a chance to adapt over evolutionarily relevant time scales. Many disastrous natural events (e.g., hurricanes, major storm waves) also tend to occur in certain intervals, enabling organisms to have developed mechanisms and behaviors to protect themselves. The human influences and the resulting changes that we bring about, however, are rapid from an evolutionary perspective and do not enable any adaptation. Of course, some catastrophes that we like to call "natural disasters" are ultimately at least partly of our own doing, such as major flooding events stemming from modifying rivers, sealing large surfaces with concrete, large-scale deforestation, or erecting settlements in areas that are prone to bearing the brunt of natural forces. Or we clearly compound the frequency and severity of certain "disasters" (e.g., forest fires). Importantly, a key human influence, namely pollution, is by definition man-made (see entry Nr. 149). Many of the chemical pollutants we discard into the environment are "novel" and confront organisms (including ourselves) with substances previously unknown in nature and against which they have no defense mechanisms. So, which is ultimately more disconcerting: natural or anthropogenic? That answer should also be clear. Ultimately, we must curb those activities that cause or aggravate the disasters that threaten our planet. It's up to us to decide whether the planet will remain inhabited by a full cornucopia of fauna and flora or, due to human disturbances, by a small range of pioneer or tolerant species along with resistant, mostly "lower" forms of life, including bacteria and viruses.

Crutzen, P.J. 2006. The "Anthropocene". In: Ehlers, E., Krafft, T. (eds.), *Earth System Science in the Anthropocene*. Berlin, Heidelberg: Springer. https://doi.org/10.1007/3-540-26590-2_3

Walker, L.R. 2011. Integration of the study of natural and anthropogenic disturbances using severity gradients. *Austral Ecology* 36: 916–922. https://doi.org/10.1111/j.1442-9993.2011.02238.x

7. **You really want to save these ugly toads/lizards/insects/birds? You're kidding, right?**
   - Define "ugly."
   - Beauty is in the eye of the beholder.
   - They can't all be pandas.

- If we killed all the ugly people on this planet, there might not be too many of us left.
- You and I wouldn't be cast as the gorgeous love interest of a new Hollywood sweetheart either.
- Attractive, impressive, or charismatic animals might be called flagship species. Many of the protection funds all over the world go to these few conservation icons ("A-list species"). Those that play a key role in the ecosystem are known as keystone species. Many keystone and other species may be inconspicuous or not especially attractive. Some are even deemed ugly or do gross things that can put people off ("D-list species"). They tend to have a PR problem, but that doesn't make them any less important. Some of the "ugly" features, such as spines, scales, tentacles, slimy bodies, or some "disproportional" body parts, no doubt evolved to protect that species from predation or to make it a more successful predator, thus promoting its survival. Just like some basically unattractive people end up with quite impressive partners, the very feature we may consider ugly in an animal may be precisely the attribute that makes that individual especially attractive to prospective mates. That helps keep the population healthy and might actually end up promoting the "ugly" feature. Moreover, those "ugly" features may be precisely those that promote survival and enable playing a role or occupying a niche that no other "cuter" species is capable of. Of course, some species are so wacky that they benefit from what might be called the ugly-cute phenomenon. Overall, protecting nature to further our own interests and esthetics is one side of the coin, respecting and protecting it based on its own hard-fought existence over countless millennia the other. It should be an issue of (bio-)ethics, not esthetics. Biologists are often criticized for being aloof or hopeless romantics: the fact that many of the species they are interested in and seek to protect are not pretty underlines that researchers in fact have a very objective, non-romanticized stance.

Jacobs, A. 2025. The hidden talents of nature's unloved animals. *National Geographic* 248: 50–71.

Snyder, E. 2013. The ugly animal: Aesthetics, power, and animal-human relationality. *Humanimalia* 5(1): 136–168. https://doi.org/10.52537/humanimalia.9970

8. **This animal is a nuisance. We should shoot them all on the spot.**
   - Your nephew or neighbor may also be a nuisance: should they be shot too?
   - This animal has become a "nuisance" only because you and I have encroached upon its territory.

- We have lots of terms for species that we would rather not have to deal with. The designations to describe some of them include nuisance species, aliens, invasive species, introduced species, exotic species, neobiota, and pests or parasites. Although these terms may have somewhat different official definitions, human activity in one form or another is behind most of them. Virtually any animal can fit the "nuisance" bill, either individually or as a population, if they appear in unexpected places or unexpected numbers or impede some human activity or threaten your poodle or . . . . They range from small to large: bats in your attic, foxes or capybaras in town, wild boar in your neighborhood park, coatis or raccoons at your vacation site, moose regularly wandering down your street, an alligator in the golf course pond, a bear rummaging through your garbage, or elephants trampling through your crops. Some were introduced on purpose: Australia is often cited as a classical case. There, camels, for example, now compete with livestock for drinking water and are shot wholesale. Of course, how all such animals are killed is also an important aspect of animal welfare and the environment. In Africa, for example, competing or threatening wildlife is often targeted with poison. Farmers may apply pesticides to water and seeds/fruits, or ranchers may add them to carcasses used to bait predators. Often, however, the chemicals kill entirely different species.

An interesting perspective is that the organism you consider a pest may be highly appreciated, held in awe, or even revered elsewhere on the planet (lions, elephants, primates, and, yes, the camel example above). Among birds, so-called nuisance species can include quite attractive forms ranging from pigeons to the Australian white ibis. The presence and abundance of the ibis, for example, have changed significantly as their native habitat has shrunken. They are now most abundant near waste management facilities. As in so many other cases, the fault tends to lie with us – we have reduced the habitat of a species and then unwittingly perhaps also offered it a new and stable food source or shelter. You are correct in recognizing that nuisance species are a problem, not least in the context of the current biodiversity crisis (where some species may displace others) but also regarding social, economic, and political considerations (due to their impact on our daily routines). The solution is intelligent, tailored wildlife management, which, however, only rarely involves shooting on the spot. By the way, more reasonable, far-sighted management would also give plenty of people like you and me promising jobs in nature conservation far into the future. And I'm sure you fully support the heavyweight mantra "jobs, jobs, jobs!" (Ted Kennedy)

Skandrani, Z. et al. 2015. Nuisance species: Beyond the ecological perspective. *Ecological Processes* 3: 1–12. https://doi.org/10.1186/2192-1709-3-3

Willis, C.S. et al. 2024. Landscape and habitat characteristics jointly drive the urban distribution of a nuisance species. *Urban Ecosystems* 27: 2283–2297. https://doi.org/10.1007/s11252-024-01594-2

9. **These animals are competing with important livestock, and this cannot be tolerated.**
   - And who is competing with you? Should we get rid of them, too?
   - You no doubt believe in fair competition. How about making that competition truly fair (and see who wins)?
   - Studies on competition between wildlife and livestock typically focus on similar-sized species and emphasize dietary overlap. The interactions, however, are usually more complex and go beyond competition (implying a negative interaction) to include facilitation (a positive interaction). In South Africa, for example, livestock facilitated younger and smaller antelope by modifying the habitat: their foraging during the wet season produced high-quality grass regrowth. A better understanding of multiple levels of interaction can help promote the co-existence between cattle and wildlife. Importantly, wildlife such as elephants and other wild herbivores typically avoid cattle. From an enlarged context, it helps to examine which side poses the greater overall disturbance. Which side has heavily modified the habitat for ranching? Who is producing enormous amounts of dung and liquid manure (and how are they disposing of it)? Several more fundamental threats tend to come from the livestock side. Accordingly, numerous wildlife species suffer from major habitat loss due to extensive livestock farming, typically involving only a single species. And farmed animals pose an increasing threat of infectious diseases for wildlife, for example, highly pathogenic avian influenza, which originated in farmed poultry, for wild birds (see also entry Nr. 203). These are key conservation and biodiversity issues.

Stears, K., Shrader, A.M. 2020. Coexistence between wildlife and livestock is contingent on cattle density and season but not differences in body size. *PLoS One* 15(7): e0236895. https://doi.org/10.1371/journal.pone.0236895

Vallis-Fox, H. et al. 2018. Water and cattle shape habitat selection by wild herbivores at the edge of a protected area. *Animal Conservation* 21(5): 365–375. https://doi.org/10.1111/acv.12403

10. **Those predators are killing our free-ranging farm animals. We have to cull them.**
    - You mean only wildlife that eats grass is OK?
    - "Cull" is a terrible euphemism – why don't you just say "kill"? (see entry Nr. 204).
    - So, the Pink Panther is the only "big cat" that should survive?
    - The fact that predators from wolves to big cats can and do kill farm animals is a legitimate concern for farm animal keepers. Again, the problem is ultimately of our own doing because we have encroached on their habitat and have given them a potentially easier alternative prey than their normal quarry. Nonetheless, the fact that all predators play an important role in the ecosystem has prompted worldwide efforts to save and even to re-introduce them where they have been extirpated (e.g., wolves in Europe). Generally, most of the negative public sentiment against predatory wildlife is based on the fear for personal safety rather than farmer complaints. The former threat is highly improbable for most people. Many strategies are available to address the issues for farmers, including habitat modification, fencing, guard dogs, hiring shepherds or herdspeople, keeping animals in barns or sheds overnight, non-lethal methods of wildlife population control (e.g., contraceptives), and government compensation for lost animals. In Botswana, for example, ranchers reduced their losses by painting large eyes on their cattle's bums! Only rarely is the solution to kill large numbers of a predatory species.

Eklund, A. et al. 2017. Limited evidence on the effectiveness of interventions to reduce livestock predation by large carnivores. *Scientific Reports* 7: 2097. https://doi.org/10.1038/s41598-017-02323-w

Laguna, M.F. et al. 2015. Mathematical model of livestock and wildlife: Predation and competition under environmental disturbances. *Ecological Modelling* 309–310: 110–117. https://doi.org/10.1016/j.ecolmodel.2015.04.020

Radford, C. et al. 2020. Artificial eyespots on cattle reduce predation by large carnivores. *Communication Biology* 3: 430. https://doi.org/10.1038/s42003-020-01156-0

11. **It's so impressive to see the whole gigantic flock of birds take off every time I drive/fly close by/shoot my gun into the air.**
    - How would you like to jump up and run around the block every time a car drove by?
    - It's too bad they can't simply hang a "Please do not disturb" sign on the doorknob.

- Wildlife is subject to many natural stresses; we shouldn't add to those with scare tactics, unnecessary bravado or by approaching too closely. How close is close? If they have to change their behavior in any way, you have impacted them. That's why whalewatching vessels, for example, have strict approach distance limits and approach angles. Merely causing whales or dolphins to change their swimming direction or stop communicating is already an issue. How close is too close? A favorite admonition by a wildlife ranger for sea lions was: "If they have to look at you, you're too close." If animals respond by fleeing, then the issue is no longer merely a behavioral one but becomes one of energetics. Fleeing, or taking flight, is an energy-expensive activity. Each such reaction leaves the animals with less energy for other survival-related activities. Such disturbances once every few hours, for example, can already reduce the fitness of flushed birds. Finally, any animal (repeatedly) flushed from its resting site is exposed to predation and may ultimately abandon that site to seek a new, perhaps less suitable shelter (e.g., if many available spots are already occupied). It may look impressive in the many nature documentaries you see on TV, but anyone with an inkling about nature should recognize it as a no-no.

Goss-Custard, J.D. et al. 2006. Critical thresholds of disturbance by people and raptors in foraging wading birds. *Biological Conservation* 127(1): 88–97. https://doi.org/10.1016/j.biocon.2005.07.015

12. **Why should we enter into a partnership with the local residents – after all, we have all the expertise, brought all the necessary equipment, and have the means to exploit the results scientifically and commercially.**
    - It's not nice to fool the natives. They might still use poison arrows or darts, or summon up their traditional head-hunter skills.
    - Watch out, the prisons in their countries are no doubt much more awful than in yours.
    - After a long history of plundering the biological bounty of indigenous peoples or taking management decisions over their heads, modern scientific practice has recognized that no conservation management project can be successful or sustainable without consulting and including local traditional knowledge and personnel. Another catchword is "community outreach." At the basic research level, this has meant including local scientists as co-workers and co-authors in the papers published in research journals. A more important level is when commercial interests or potential profits are involved, such as when the local biodiversity is screened for promising active substances with

pharmaceutical promise or for genetic resources (for example, 80% of registered medicines come from plants or have been inspired by natural products). This has given rise to new agreements such as the Nagoya Protocol (Nagoya Protocol on Access to Genetic Resources and the Fair and Equitable Sharing of Benefits Arising from the Utilization) (Access and Benefit Sharing: ABS)). That protocol is a supplementary agreement to the Convention on Biological Diversity (CBD). This legal framework was adopted in 2010 and entered into force in 2014. The take-home message is twofold: (1) the plants and animals that could help cure diseases such as cancer could at this very moment be going extinct without cooperative nature conservation efforts, and (2) the residents/peoples at the exploited sites must be involved in and benefit from the commercial exploitation of local biodiversity. Importantly, traditional knowledge can enable an equitable exchange of information, foster a shared responsibility, and lead researchers more quickly on the right path.

Drew, J.A. 2005. Use of traditional ecological knowledge in marine conservation. *Conservation Biology* 19: 1286–1293. https://doi.org/10.1111/j.1523-1739.2005.00158

https://www.cbd.int/abs/about

https://www.cbd.int/article/pharmaceuticals-biodiversity-planet

13. **Yes, this mega-project will attract thousands of tourists, but each will pay a small fee that will be used for nature conservation, so it's a win/win situation.**
    - Name me a tourist fee you ever paid that went directly to support a conservation project.
    - And the whopping commercial profits will go where exactly?
    - The emphasis is usually on the negative impacts that tourism development inflicts on the environmental resources on which it depends. Is there some positive conservation-related feedback? Theoretically, or let's say potentially, yes. Direct options include revenue from park-entrance fees or fees from conservation-related activities (e.g., night-time sea turtle nesting beach tours). Visitors who have enjoyed nature-based tourism are also sometimes willing to make (later) donations. Funds generated by the rental of recreational equipment or from hunting/fishing licensing fees can help manage natural resources or pay for park ranger salaries. More indirectly, some governments might use the respective tax revenues for environmentally relevant initiatives. Nonetheless, specific or more broadly valid quantitative data on any such cash flows are rare. The arguments for specific fees are typically related more

to minimizing the damage by visitors (e.g., funds for fencing), ensuring visitor comfort, or feigning sustainability. Examples include tangential clichés such as energy-efficient resort infrastructures, modern waste treatment and disposal facilities, or "an increased public appreciation for the environment and awareness for environmental problems" (for example, by erecting signage with "green" messages). Such funds only rarely bring about a perceived direct positive influence of tourism. One constructive approach, practiced for example by certain resorts in the Maldives, is to build an adjoining bio- and info-station and hire a resident biologist to hold lectures and accompany guests on snorkeling and diving tours. Beyond helping to minimize destructive human activities in the delicate reef ecosystem, it also raises a broader environmental awareness.

Hehir, C. et al. 2022. Last chance for wildlife: Making tourism count for conservation. *Journal of Sustainable Tourism* 31(5): 1271–1291. https://doi.org/10.1080/09669582.2022.204980

Kikinger, R. 2003. Approaches to a nature-based management in marine tourism. In. Bright, M., Dworschak, P.C., Stachowitsch, M. (eds.), *The Vienna School of Marine Biology: A tribute to Jörg Ott* (pp. 113–120). Wien: Facultas Universitätsverlag.

Sunl, U. 2003. Environmental impacts of tourism. In: Camarda, D., Grassini, L. (eds.), *Local Resources and Global Trades: Environments and Agriculture in the Mediterranean Region*. CIHEAM, Options Mediterraneenes: Serie A. Seminaires Mediterraneens, n. 57: 263–270. https://www.researchgate.net/publication/275349611_Environmental_impacts_of_tourism

14. **The changes we plan to make to this beach/riverbank/reef, etc. will make it much more beautiful than it was originally.**
    - Define "beautiful."
    - Did he actually say "Make America beautiful again"?
    - Beauty is a matter of fashion. Do you want nature to become part of the fickle fashion industry?
    - It's true that the beauty of nature is a major driver for visitors. Beauty, however, is always in the eye of the beholder. Dozens of factors may be involved in determining the subjective impression of beauty (e.g., scale, time, condition, sound, balance, diversity, novelty, shape, uniqueness – even smell). This is the subject matter in the field of environmental psychology. Some people are blown away by an endless desert, others by Arctic glaciers, and the two factions would probably never come to an agreement on which is the more beautiful. Other so-called beautiful habitats can actually be an endless source of threats and

unpleasantries – better savored lying on a couch and watching TV documentaries: few simply equipped and clad people would ever really enjoy spending several days and nights in a rainforest.

The state of the natural environment is the result of the long-term effect of the forces of nature (including geology, meteorology, etc.) and the interactions between the environment and the fauna and flora that occupy it (that's the basic definition of "ecology"). Changing that status or configuration through our actions is an interference in natural processes. And the motivation to interfere is almost always greed, not the desire to improve beauty. Importantly, such artificial interventions typically upend a certain balance that nature tends to try to re-establish. Working against nature and maintaining any such man-made "beauty" is an expensive proposition that always requires a major input of energy and resources. On a smaller, simpler scale, think about the effort you need to maintain your well-manicured lawn or garden. Those wishing to capture and cash in on the beauty and esthetics of nature are more likely to be successful over the long term by leaving things as they are.

Kirillova, K. et al. 2014. What makes a destination beautiful? Dimensions of tourist aesthetic judgment. *Tourism Management* 42: 282–293. https://doi.org/10.1016/j.tourman.2013.12.006

15. **Biodiversity? Isn't that the dishwashing liquid with the frog on the label?**
    - If you believe that, then you'll believe that storks bring babies.
    - So, you think this "diversity" thing has really gone too far, yet another "woke" issue in need of backtracking?
    - People really can't be blamed for being baffled by the emphasis on biodiversity, a term that wasn't even in any dictionary not too long ago. From the nature conservation or species protection perspective, the notion can be more simply understood as the complete fauna and flora of a particular area. One benchmark in gauging biodiversity is the difference between the organisms that were there historically and those that remain currently after human impacts and interventions. Practically, it helps to envision each organism in the habitat as being equivalent to a rivet in the fuselage and wings of an airplane. Like each rivet, every organism plays a role in the functioning of the overall system. Removing a rivet (i.e., making one species go extinct) may have no noticeable effect, with another rivet (or species) potentially taking over the function of the others. At some point, however, removing one

rivet too many will cause the airplane to disintegrate in full flight (the rivet popper hypothesis). Our information about ecosystems is by no means as sophisticated as our knowledge about airplane construction, meaning that we are unable to predict which species and when their removal may damage the entire ecosystem. The reasonable strategy in such cases might be to invoke the precautionary principle, which in simplified form states that if we don't know exactly what we are doing, it's better to be safe than sorry.

Ehrlich, P., Walker, B. 1998. Rivets and redundancy. *BioScience* 48(5): 387. https://doi.org/10.2307/1313377.387

https://medium.com/@say4mitalig/the-rivet-popper-hypothesis-93a940f44751

16. **Nature is really brutal (mosquitoes, diseases, predators, bacteria, hurricanes), so we have to tame it to the max.**
    - "It's not nice to fool Mother Nature." (Dena Dietrich, 1970s TV commercial)
    - So, the top priority on this planet is your comfort?
    - On the one hand, pristine nature – and often even compromised semi-natural or moderately damaged ecosystems – seems beautiful or esthetic to us. On the other hand, humankind has always been challenged to understand and subdue nature, and to make it "useful." Why? Because we feel threatened by it: any intact ecosystem involves a never-ending web of eat and be eaten. And we don't really want to be directly part of that (at least not the second, "eaten" aspect). Whether the food chain involves a top predator crushing a poor herbivore or a bacterium killing off its host – it's never a pretty sight. Human comfort was never a consideration in the natural order of things (see entry Nr. 180 for dominion over nature). From this perspective, nature conservationists and those trained in ecological matters (or naive-romantic "tree huggers" as some may call us) are actually more objective and rational than those who wish to bend or tame nature to their own well-being or commercial interests. So maybe it pays to listen to and follow the advice of trained, conservation-minded folk. Today, the answer must be a compromise between untamed nature and human survival, whereby our survival looks bleak if the world's ecosystems cease to function.

Melville, G., Ruta, C.R. 2019. *Nature and Human: An Intricate Mutuality*. Berlin, Boston: De Gruyter Oldenbourg. https://doi.org/10.1515/9783110579765

17. **In the future, advanced technologies will solve all these problems – and we need more data before acting now anyway.**
   - So, you want to pass your sins off to a future generation to solve?
   - When was the last time high-tech solved a fundamental environmental or other deeply human problem?
   - Technology may allow us to bend the rules, but only a bit and not for long.
   - Yes, advanced technologies can go a long way toward recognizing and visualizing certain environmental issues (e.g., satellite imagery of declining forest cover) and to help solve some of them (modern wastewater treatment). Such technologies, however, are often one step behind the environmental deterioration they document and seek to ameliorate. Importantly, many problems and threats must be solved today before irreparable harm is done. Ecosystems that are lost are not necessarily restorable. No technology will revive extinct species (even if certain high-tech companies are considering it under the term "de-extinction"; see entry Nr. 26). Importantly, waiting to address problems in the future, when conditions have worsened even further, will always be considerably more expensive and time-consuming than dealing with them at an early stage.

   One commonly heard excuse for not taking immediate action is the hollow and counterproductive mantra that more data or information are required. This is a fatal misconception because, in most cases, we will never (or only some time far in the future) fully understand how ecosystems function and potentially be in a position to micromanage a problem. Instead, we should rely more fully on common sense and basic conservation guidelines today. That common sense tells us even now that straightening rivers, overfishing the seas, polluting the environment, large-scale felling of pristine forests, and killing species that are a perceived nuisance or pose any kind of threat to anything is a recipe for disaster. And, speaking of technology, I would hope you are not placing your bets on megalomaniac ocean-/solar-/climate-/or stratospheric geoengineering schemes that promise sci-fi technological solutions to burgeoning environmental problems. Or maybe you are setting your hopes on nuclear fusion, perhaps on the first commercial mega-pilot facility in your neck of the woods? Or are you relying on your family escaping the problems on Earth by colonizing Mars?

Mast, R. 1999. Common sense conservation. *Marine Turtle Newsletter* 83: 3–7. http://www.seaturtle.org/mtn/archives/mtn83/mtn83p3.shtml

18. **This environmental disaster can't be that bad because it has happened elsewhere before and you never read about it again in the media (Figure 1).**

Figure 1 Not many of the thousands of pollution events yearly make their way into the news. Even if only locals heard about it, this small beach in California was tarred nonetheless and required a professional cleanup.

- That's because obituaries don't tend to be repeated or updated at regular intervals.
- Just because you don't read regular updates about a major environmental impact doesn't mean that the damage has subsided or that the affected ecosystems have recovered. Of course, this depends heavily on what you read. The daily news, as the name implies, always focuses on the current events and "situation room"-type stories. Most manmade ecological disasters quickly disappear from the headlines, with events of past years and subsequent (environmental) recoveries rarely receiving further mention – unless the economic, social, or political repercussions of the damage remain disastrous or close enough that they cannot be overlooked. In fact, papers on major environmental disasters and their further consequences *are* regularly published in the respective scientific journals long after they occurred. Accordingly, papers with updates on major oil spills, radioactive contamination, and toxic chemical accidents continue to appear years and even decades later. Yet another service that researchers provide.

Brian Miles, B., Morse, S. 2007. The role of news media in natural disaster risk and recovery. *Ecological Economics* 63 (2–3): 365–373. https://doi.org/10.1016/j.ecolecon.2006.08.007

Lunney, D., Moon, C. 2012. Disasters for wildlife: An analysis of media attention. *Australian Zoologist* 36(1): 5–19. https://doi.org/10.7882/AZ.2012.002

19. **No matter what we do, nature is powerful and will recover completely.**
    - So, if pollution withers your garden, your flowers will magically reappear unscathed next year?
    - Maybe nature can recover if the trolls and hobbits do their share?
    - That depends a bit on how you define "powerful" and "nature." None of the organisms that have been driven to extinction by humans were powerful enough to escape that fate or will ever have the power to "recover" or reappear on our planet. That goes beyond the dodo, the Tasmanian wolf, Steller's sea cow and passenger pigeons to include untold scores of less conspicuous species, many of which will never even be described (but see entry Nr. 26). Is there something worse than being dead? Yes. Being extinct. The current rate of human-influenced extinction is estimated to be 1,000 times that of the background rate, with an estimated 1 million of the planet's plant and animal species at risk of extinction.

    The situation does not necessarily look any better for entire habitats or ecosystems. When disturbed, they can at best undergo long-term recovery in the form of a succession involving initial pioneer forms and a sequence of intermediate development or seral stages. These, however, do not necessarily lead to the original natural (climax) stage. Non-reversible habitat loss processes include desertification, large-scale deforestation through slash-and-burn procedures, or major losses due to erosion. Equally problematic but sometimes less conspicuous changes can be referred to as ecosystem shifts, which typically involve relatively sudden and drastic rather than gradual, evolutionary changes. Many such shifts – depending on the causes – are also poorly reversible. Importantly, restoring damaged ecosystems requires considerably more time, effort, and money to correct than to prevent in the first place. This calls for stepping up now.

Pimm, S.L. et al. 2014. The biodiversity of species and their rates of extinction, distribution, and protection. *Science* 344(6187). https://doi.org/10.1126/science.1246752

Scheffer, M. et al. 2001. Catastrophic shifts in ecosystems. *Nature* 413: 591–596. https://doi.org/10.1038/35098000

2
Science

20. **You come here, you collect the specimens and data, you leave and publish by yourself and then commercially exploit the results. We are left with nothing.**
    - I'm truly sorry for the misdeeds of the past, but let me prove you wrong in our case.
    - As a biologist I can assure you I myself am not earning any serious money by being here.
    - It's true that researchers in the past have often behaved that way ("colonial science," "research colonialism," "data colonialism"). Today, however, that is generally no longer the case and everyone is required to follow strict guidelines covering all the above issues. It begins with project submission. A key point in submissions for funding by national and international agencies is the inclusion of cooperation partners. Major EU projects, for example, call for the inclusion of researchers from several countries, whereby there is often a specific requirement to include partners from developing countries. Such partners are then involved in all phases of the research, from fieldwork to publication, with training of local scientists typically being an integral feature. In nature conservation projects, community outreach is a major cornerstone. Moreover, no research or collections are possible today without official approval and permits from the responsible agencies of the respective countries (see also entry Nr. 12).

Adams, W.B., Mulligan, M. (eds.). 2003. *Decolonizing Nature: Strategies for Conservation in a Post-Colonial Era.* Routledge. 320 pp. https://doi.org/10.4324/9781849770927

https://research-and-innovation.ec.europa.eu/strategy/strategy-research-and-innovation/europe-world/international-cooperation_en

DOI: 10.1201/9781003757474-2

21. **For every study or expert opinion asserting one thing, I can bring you another one saying the opposite.**
    - "Our heads are round so our thoughts can change directions" (Francis Picabia).
    - Are you referring to results published in recognized scientific journals or listening to bloggers and influencers?
    - A fundamental process of science is to develop hypotheses and theories to explain natural phenomena and then to test these by observations and experiments. Probably every natural phenomenon has had two or more competing underlying theories, often with heavyweight proponents on each side. Many ultimately correct theories were initially disparaged and reviled both by eminent scientists and by the public. A wonderful feature of science and scientists, however, is that once a theory is proven wrong, then researchers are typically the first to discard that falsified theory. In some cases, the "battle of the theories" and the quest for truth can take decades or longer. That battle is often unnecessarily protracted by personal sensitivities and prejudices, commercial interests, and by political and/or religious standpoints. Unfortunately, some scientifically tried-and-tested, indisputable theories will continue to be denied by certain segments of the public based on such standpoints. This can even manifest itself in a more general "anti-science" ideology. In nature conservation or health matters, for example, it is often informative to determine whether and what vested interests may be behind iffy environment-related expertises and websites (the tobacco industry? oil & gas corporations? auto makers?). Today, fake news and alternate facts tend to further confuse and mislead the general public. Take the time to look behind the hype – you don't want to be on the wrong side when the truth ultimately prevails.

Caudill, D.S. 2023. *Expertise in Crisis: The Ideological Contours of Public Scientific Controversies* (Bristol, 2023; online edn, Policy Press Scholarship Online, 21 Sept. 2023), https://doi.org/10.1332/policypress/9781529230918.001.0001

How science works. https://undsci.berkeley.edu›uploads›2022/09

22. **You researchers certainly kill a lot of animals.**
    - But the goal is to save a lot more.
    - Who kills more wildlife: researchers or hunters/poachers, commercial fisheries (including bycatch), industrial agriculture, large-scale forestry operations, etc.?
    - Traditionally, many specimens were killed by researchers for museum collections, for educational purposes (lab courses), or to study animal

community compositions (for example, by dredging the seafloor). Today, non-lethal research is gaining the upper hand in all these activities. Many lab courses that once required killing and dissecting numerous specimens are now held using virtual, digital methods (no scalpels necessary). Approval from bioethics commissions must be sought when submitting research projects that plan to kill or compromise the welfare of organisms. A reference to the code of practice adopted for the reported experimentation or manipulation and permit numbers must be listed in all publications that present such research results. Journal editors and reviewers are required to reject papers where such information is missing. Modern statistical methods are also enabling solid interpretations without disproportionate sampling. In research on whales, for example, lethal methods – such as those once almost exclusively used in the framework of scientific whaling ("whaling under special permit") under the auspices of the International Whaling Commission (IWC) – can be almost entirely replaced by a wealth of non-lethal methods such as visual surveys (transects), photo-IDs, genetic approaches (skin biopsies), acoustic studies, tagging, examining stranded animals, feces analyses, and whale watching.

Franco, N.H. 2016. 76. Killing of animals in science–is it always inevitable? *In: Food Futures: Ethics, Science and Culture* (pp. 499–504). Wageningen Academic.

https://iwc.int/resources/protocols-guidelines-and-how-to-guides

https://iwc.int/scientific-research

3
Extinction and Evolution

23. **Many species have died out in the past, and the world is still standing: why worry about this particular species now?**
    - Is the world really still standing tall (or already hunched)?
    - Each species is like a rivet holding a plane together. Maybe losing this one will trigger the catastrophe in mid-flight?
    - It's true that many species have died out in the past and, in fact, five major mass extinctions have been recorded in Earth's history. Almost all these extinctions occurred over millions of years, that is, "gradually" and not at the current breakneck speed. The situation today is being described as the "6th major extinction event on the planet." Many of the species currently going extinct are inconspicuous or even as yet undescribed. Each plays a role in the ecosystem, and its loss can destabilize that system (see rivet analogy in entry Nr. 15 under "General"). This is all the more reason to focus on saving those that we do recognize and to avoid the large-scale destruction of habitats that no doubt harbor untold undescribed species (think rainforest). Often, protecting a recognized species is a major step in saving those less-recognized or less-recognizable species in the same habitat. This is one important aspect of the "flagship" species concept: protecting that iconic animal can go a long way toward helping those around it that we know little or nothing about. Accordingly, beyond concentrating solely on helping a particular species (species protection), nature conservation and wildlife management will typically involve protecting the respective habitat and "managing" local stakeholders.

    Species extinction: Who cares? Why? https://laurencerose.co.uk/2020/09/extinction-who-cares/.

    https://laurencerose.co.uk/2020/09/extinction-who-cares/

DOI: 10.1201/9781003757474-3

24. **The dinosaurs also died out, and we sure are all the better for it.**
   - But that was cosmic doing, not human greed.
   - Yeah, but they would have made a better guard dog than your Rottweiler.
   - True, it's hard to imagine what kind of lives we would be living if dinosaurs still roamed the Earth. Maybe we would have survived long enough to develop weapons that would have enabled us to kill them all, much like we did with certain large mammals in early human history. In comparison, the few big cats, wolves, and other top predators still alive today would have been a very minor irritation. One flippant take on the dinosaurs – "The Age of Reptiles ended because it had gone on long enough and it was all a mistake in the first place" (Will Cuppy) – would be an interesting thought if applied to humans. When the dinos died out, however, so did much other life on the planet – and this probably wasn't a pretty sight. Back then, it was a cosmic event (asteroid impact). And it took millions of years more for new animal species and communities to evolve to take their place.

     Do you want that to happen again, this time due to human ignorance and complacency? This means that the issue can also be viewed from an (bio-)ethical perspective: do we humans have the right to decide which species have to go and then actively or passively make that happen? Natural events that can kill off entire populations or even species will continue to occur (for example, on islands due to volcanic eruptions or major storms), and there is little we can do to intervene there. We should therefore concentrate on cases where action today (or simply leaving nature alone) can prevent species losses. This common-sense approach is valid even if we don't yet exactly know a particular species' role in the ecosystem (ecosystem function) and, ultimately, what it does for us (ecosystem services). Extinction is forever.

Devereaux, A. et al. 2004. Are dinosaurs the rule or the exception? Developing concepts of death and extinction. *Cognitive Development* 19(3): 363–383. https://doi.org/10.1016/j.cogdev.2004.04.001

25. **If this species can't survive on its own without our help, it probably doesn't deserve to survive.**
   - So, if your grandmother can't cross the street on her own, she has no right to live?
   - We're not asking you to help, just to stop hurting.
   - Most species aren't threatened with extinction because they are poorly adapted to the environment. They wouldn't be here if they weren't. Accordingly, it's not a lack of fitness. Moreover, "survival of the fittest"

is inapplicable in today's highly altered and managed environments – unless we wish to remain surrounded solely by highly resilient species such as rats. Nor can our conservation efforts be seen as hopeless attempts to painstakingly maintain or restore a precarious species composition in a highly compromised environment. Rather, we altered their habitat and should do everything we can to make things right. In an age when human activities are shaping the face of the Earth ("Anthropocene"), we are responsible for the damage and for rectifying it. And it's ultimately in our own interests because living in impoverished ecosystems denies us and future generations of basic human needs and rights. Nature conservation can be something that can make us both feel good and feel better (healthwise), not to mention help guarantee our very survival.

Milner-Gulland, E.J. et al. 2014. Accounting for the impact of conservation on human well-being. *Conservation Biology* 28(5): 1160–1166. https://doi.org/10.1111/cobi.12277

26: If this species becomes extinct, scientists will be able to re-create it by genetic means.

- Well, that idea almost made the owners of Jurassic Park rich, right?
- Maybe in a very few cases we can come close, but it would be thousands of times more expensive than saving the species now. The term for recreating a species is "de-extinction." One company involved in this issue defines it as:

"The process of generating an organism that both resembles and is genetically similar to an extinct species by resurrecting its lost lineage of core genes; engineering natural resistances; and enhancing adaptability that will allow it to thrive in today's environment of climate change, dwindling resources, disease and human interference."

Note the words "resembles" and "similar" – not "the same." Moreover, artificially engineering enhancements and resilience into such "new" organisms is apparently part of the plan. Does that include resilience to poaching? In 2016, the International Union for the Conservation of Nature (IUCN) provided a definition, which was deemed faulty and corrected by Novak (2018) to:

"De-extinction is the ecological replacement of an extinct species by means of purposefully adapting a living organism to serve the ecological function of the extinct species by altering phenotypes through means of various breeding techniques, including artificial selection, back-breeding and precise hybridization facilitated by genome editing. The goal of de-extinction is to restore vital ecological functions that sustain dynamic processes producing resilient ecosystems and increasing biodiversity and bioabundance."

Currently, several species – among them the woolly mammoth and Tasmanian tiger – have been singled out as potential candidates for de-extinction. Each effort may, if ever successful, take decades to realize and cost many millions. One company has apparently already bred three pups that carry several genes of the extinct giant dire wolf; that involved creating embryos from edited gray wolf cells and using surrogate dog mothers. In light of the estimated 30,000 species being driven to extinction each year ("6 per hour") and many other dire predictions, can de-extinction be the way to "ensure a brighter tomorrow" and "a more sustainable future," as one company writes? You decide.

https://colossal.com/de-extinction/

Novak, B.J. 2018. De-extinction. *Genes* 9(11): 548. https://doi.org/10.3390/genes9110548

27. **Only the strong(est) survive. So why don't we just wait and see who wins?**
    - So, you would be willing to fight to the death with your co-workers and see who keeps their job?
    - It certainly sounds convincing when Billy Paul ("Only the strong survive . . . the weak fall by the wayside") sings it.
    - That saying, by the way, was apparently never stated in this manner by Charles Darwin himself, so those using it are not quoting that icon. Moreover, it's not actually about being physically strong; it's more about being best adapted to the environment. "Survival of the fittest" is a related phrase (by Herbert Spencer, 1864, not Darwin). While perhaps at least an apt point of discussion in connection with animals, its application to humans ("social Darwinism") in one form or another is terribly misguided and has led to some of the most horrific atrocities in recent human history ("inferior races," "unworthy of life," . . .). The general notion, however, can be related to the theory of natural selection, whereby the emphasis is on "natural." The vast majority of species *are* optimally adapted to their environment – otherwise they wouldn't be here. Or better said, they at least *were* highly adapted to their environment before humans began to exploit and alter the planet in earnest.

    Of course, it's a basic tenet of ecology that the environment influences organisms and that organisms, in turn, influence their environment. Those changes and the relevant adaptations are typically long-term, evolutionary processes that take place over many millennia to millions of years. While that's only a blip in geological history, it's still excruciatingly slow compared to the rates and magnitudes of environmental changes that humans have initiated. Most species simply cannot adapt (i.e., genetically) that quickly. Those that adapt quickly behaviorally, rather than receiving our recognition or admiration, are often stamped as being a threat or a nuisance (e.g., rats). Those that do survive major disturbances, such as pollution events, are typically less sensitive, so-called lower life forms. We can decide whether we wish to live in and fight for a biodiversity-rich environment or be surrounded by an impoverished set of hardy survivors, ultimately on the level of certain "simpler" invertebrates (like worms) along with bacteria & Co.

Claeys, G. 2000. The "survival of the fittest" and the origins of social Darwinism. *Journal of the History of Ideas* 61(2): 223–240. Project MUSE, https://doi.org/10.1353/jhi.2000.0014. https://muse.jhu.edu/article/15068

28. **OK, so this habitat is polluted, but the animals will evolve and adapt to it.**
    - Well, if we humans can't evolve and adapt to radioactivity, heavy metals and the like, why should animals?
    - If you're dead, you can't evolve (except gas maybe . . .).
    - Pollution, which is defined as being caused by humans (and does not include "natural" events such as volcanic eruptions), is a brand-new phenomenon because humans have been on this planet for only a blink of the eye, geologically speaking. Since evolution is a long-term process, no short-term adaptation based on evolution is possible. Of course, regardless of the severity of a pollution event or chronic pollution, some organisms always do seem to still be around. This is less the result of (genetic) adaptation than it is of (1) being particularly insensitive to a certain pollutant or (2) being the first organisms to return after pollutant levels have dropped to survivable levels. In general, the lower a species is on the "complexity" totem pole, the less sensitive it is (for example, to radioactivity). Accordingly, bacteria and certain "lower" invertebrates are much more likely to survive than mammals. And humans may be the most sensitive of all. So, if we are disinclined to reduce pollution for the sake of maintaining biodiversity and intact ecosystems, then we certainly should take action to protect humans occupying the same environment.

    Of course, some pollutants are so toxic that no organisms can survive prolonged exposure. Many of those chemical compounds are synthetic and were never before present in nature (e.g., persistent organic pollutants – POPs), and many were actually expressly designed to kill organisms such as insects in the first place. Other toxic substances are present inside the planet (geologically) but are not part of biological systems (e.g., mercury). And others are lethal because of their "dose"

("Sola dosis fecit venenum" – only the dose determines whether something is toxic or not. Salt, for example, is nice to have on the dinner table, but anything more than a few tablespoons at once and you are toast). Death is, of course, only one response to pollution. The list of sublethal physiological, neurological, and behavioral effects in animals and humans is long and unpleasant. In humans: profuse sweating, sleeplessness, compromised growth and fertility, chromosome damage, shorter lifespan, cancer. What, you mean you're sweating more than usual and aren't sleeping well these days!?

Loria, A. et al. 2019. Mixed evidence for adaptation to environmental pollution. *Evolutionary Applications* 12(7): 1259–1273. https://doi.org/10.1111/eva.12782

29. **No one really knows or cares about that stupid species.**
    - "Only if we understand, can we care. Only if we care, we will help. Only if we help, we shall be saved." (Jane Goodall)
    - "In the end, we will conserve only what we love; we will love only what we understand and we will understand only what we are taught." (Baba Doium)
    - "We are all ignorant; just about different things." (Mark Twain)
    - Each one of us probably knows or can name only very few species. Nobel Prize–winner Konrad Lorenz once bemoaned that youngsters were able to recognize a dozen car makes but could identify only very few bird species. Today, there are an estimated 8–10 million living species, of which only a bit more than 1 million have been scientifically described. Of those, the conservation status of only about 170,000 has been assessed by the International Union for Conservation of Nature (IUCN), whereby 46,000 are listed as being threatened with extinction. Most of these involve larger and more conspicuous mammals, birds, reptiles, amphibians, and fish that we and our children love to watch on TV documentaries. Of course, according to some estimates, 45% of flowering plant species are also potentially threatened with extinction. Importantly, even if we personally can identify only a handful of species, and the IUCN has had time to assess the status of only a fraction of them, they all are out there and all playing a role in the ecosystem (not to mention a role in making our lives possible and sometimes even enjoyable). Each one is the product of millions of years of evolution and, as such, deserves our respect and, when necessary, our help. Beyond purely biological/ecological considerations, this issue can also be discussed from an ethical or moral standpoint and can culminate in something known as species egalitarianism (the view

that all species have equal moral standing and should, at minimum, command respect). That viewpoint would require taking the interests of all so-called non-human animals into account. . . .

https://www.iucnredlist.org/

Schmidtz, D. 2002. Are all species equal? *Journal of Applied Philosophy* 15(1): 1–121. https://doi.org/10.1111/1468-5930.00073

30. **That species is already almost extinct. It's too late to help them.**
    - Better late than never.
    - "There is an immeasurable distance between late and too late." (Og Mandino)
    - It's never too late. If the remaining individuals in the wild include a few mature females (or if captive animals are also available), then the species, if fully protected, can still survive. The time to act is now. Later, no amount of money or effort can bring them back. Allowing or accepting extinction is also clearly contrary to the fundamental tenets of conservation biology. It also runs counter to broader ethical and moral considerations.

    Among the porpoises, for example, after a precipitous decline in recent decades, only a few individuals of the vaquita remain in the Gulf of California. Is there still hope? Yes. In fact, saving the vaquita should be a no-brainer: a single, clearly defined species, in a single, small, clearly defined protected zone, with one clearly recognizable threat (bycatch in illegal fishing nets for an endangered fish species) emanating from only two to three villages in Mexico. Of course, this also requires a wider, concerted effort by scientists, NGOs, politicians, and the governments of Mexico, China (to which the swimbladders of the endangered fish are exported), and the USA (transit route for the swimbladders). Major targeted action earlier could have saved ongoing major expenses and fire-brigade-like rearguard actions. It would also have helped save face for the main country involved.

Bessesen, B. 2018. *Vaquita. Science, Politics, and Crime in the Sea of Cortez.* Island Press. 293 pp.

Jachowski, D.S., Kesler, D.C. 2009. Allowing extinction: Should we let species go? *Proceedings of the National Academy of Sciences USA* 105(4): 2919–2922.

Jaramillo-Legorreta, A. et al. 2007. Saving the vaquita: Immediate action, not more data. *Conservation Biology* 21(6): 1653–1655. https://doi.org/10.111/j.1523-1739.2007.00825.x

4
Terrestrial

31. **Bats are creepy, bloodsucking pests and I have kids. Call the exterminator.**
    - I thought you liked all the Batman movies?
    - So, you're more a Joker and Scarecrow fan?
    - Only 3 out of the more than 1,400 bat species worldwide feed on blood, and they are entirely absent in Europe and Asia. Of those three, the hosts of only 1 (the common vampire bat) are large mammals – the other two feed on birds. We have indirectly contributed to the success of the common vampire bat due to wide-ranging cattle farming. Choosing a human to bite is a true rarity for that species. Efforts are being made to reduce its population, but this reflects its role in spreading rabies. In fact, you can be thankful for what most bats do for you day and night (mostly night, of course). They play major roles as pollinators (agave, saguaro, (wild) banana, durian) and help scatter the seeds of nuts, figs, cacao, and wild pepper. Some plants, such as the agave (think tequila!) and saguaro cactus, would apparently go extinct without bats. And if you have to hate a species, maybe choose mosquitoes. In that case, you should know that a single bat can consume up to 1,000 of those pests in one night (some of which are human disease vectors)! This translates into tons of insects nightly for a large colony and thus also a key role in eliminating crop and forest pests. Finally, their droppings (guano) are a highly sought-after and expensive fertilizer – maybe you can cash in on "your" bats (but wear a mask to avoid inhaling the dust of any animal droppings).

Greenhall, A.M. 2018. *Natural History of Vampire Bats*. Boca Raton: CRC Press. 245 pp.

https://www.si.edu/spotlight/bats/batfact

Tuttle, M. 2018. *The Secret Lives of Bats: My Adventures with the World's Most Misunderstood Mammals*. Mariner Books. 304 pp.

DOI: 10.1201/9781003757474-4

32. **There are so many frog species here, surely losing a few can't be a real problem.**
    - Losing how many would you say is OK?
    - You're only saying that because you're the biggest frog in the puddle.
    - The important thing to remember is that each frog species plays a different role in the ecosystem, even if we often don't yet know exactly what all those roles are. This diversity is perhaps reflected in an extraordinary range of elaborate social behaviors, especially in connection with territoriality, courtship, mating, and parental care. Some lay their eggs in ponds, others in the rainwater trapped between plant leaves, others high up on tree leaves (so that the larvae drop into the water below), and still others on their own backs. A good question to ask is why frog populations are threatened. Could it be due to human activity? There you go! Importantly, some groups such as the so-called poison or poison dart frogs are of direct benefit to indigenous tribes and have attracted the attention of researchers because the toxins present an opportunity for new drug discoveries. And don't forget, extinction is forever – once a frog species is lost – and there are many yet to be described, and some occupy very restricted habitats – we can't bring them back. Of course, the same logic used above can be applied to lizards, birds, mammals – do we really "need" them all?

Summers, K., Tumulty, J. 2014. Chapter 11 – Parental care, sexual selection, and mating systems in neotropical poison frogs. In: Macedo, R.H., Glauco, M. (eds.), *Sexual Selection* (pp. 289–320). Academic Press. https://doi.org/10.1016/B978-0-12-416028-6.00011-6

33. **This is not a pristine (old-growth/virgin/primeval) forest, so nothing much is really lost if it's cut down.**
    - So, you want to degrade this already degraded forest further to scrubland or fallow?
    - Even second-growth or semi-natural forests are valuable habitats. This can be viewed from two angles. If we had a green light to cut down everything that was already modified or non-pristine, there wouldn't be much nature left on this planet: that would be carte blanche to cut down the entire forests of most developed countries (most of which are "secondary" rather than "primary"). Conversely, if only pristine forests were valuable and we were called upon to protect only them, then we would be in a real quandary. Conservationists would have relatively little to do in many countries because the virgin or first-growth forests in many have long been logged or slashed and burned. Recovering forests often represent some stage in the succession of vegetation communities, ultimately leading to a climax community that may boast a

composition very similar to that of the original old-growth or primary forest. That process may take anywhere between one and three centuries. Such forests typically offer the full range of vertical forest levels (e.g., from understory layer to canopy) along with the structures and ecological niches that enable the highest biodiversity. These are all valuable arguments for forest restoration schemes. One overarching perspective involves asking what a pristine forest actually looks like and whether any actually (still) exist (yes). Another is to ask whether we are capable of recognizing the pristine state if only a few of us have ever experienced it. Most of us can't because most of us haven't (see "the shifting baseline syndrome" in entry Nr, 163).

Barlow, J. et al. 2012. How pristine are tropical forests? An ecological perspective on the pre-Columbian human footprint in Amazonia and implications for contemporary conservation. *Biological Conservation* 151(1): 45–49. https://doi.org/10.1016/j.biocon.2011.10.013

Buchwald, E. 2005. A hierarchical terminology for more or less natural forests in relation to sustainable management and biodiversity conservation. https://www.fao.org/4/Y4171E/Y4171E54.htm

Sayer, J. et al. 2004. The restoration of forest biodiversity and ecological values. *Forest Ecology and Management* 201(1): 3–11. https://doi.org/10.1016/j.foreco.2004.06.008

34. **This plant has such beautiful flowers that I brought back a bag of seeds with me from vacation and am planting them everywhere to brighten up the neighborhood.**
    - Sounds like material for a good movie title: "Surrounded by aliens."
    - Plants are not vacation souvenirs.
    - This is a topic that "opens up a whole can of worms." Overall, introducing exotic species is a major conservation issue, with most cases being evaluated negatively from a variety of standpoints. These include competition with native species, the potential to become invasive and uncontrollable, the questionable value of exotic plants to native invertebrates and herbivores, the fewer insects that are adapted to pollinate the particular species, the introduction of parasites, and the potential toxicity to animals and humans. These issues are standard fare for ecosystem managers but are often provoked by folks who are blissfully unaware of what they are potentially unleashing.

    Hobby gardeners are a classic example, whereby the two extremes are "the garden of Eden" and "the spick-and-span garden":
    – the former are those who may simply wish to design a backyard that is absolutely unique and eye-catching. It's easy to succumb

to the temptation to select the most colorful species possible, regardless of their original native habitat. These should, wherever possible, be side-stepped, even if legally imported or raised locally. Professional garden center staff will no doubt be able to suggest native plants that fill the bill (and that promise less care and longer-term survival) and

– the latter are those who rigorously strive for an ultra-tidy garden: the grass is regularly closely cropped, no fallen leaf or "weed" is allowed to mar the manicured surface, pesticides/herbicides are applied to keep "pests" away ("hey, why can't I use pesticides in my own garden – everything is artificial anyway"). Here, neighbors with unruly gardens are looked upon disdainfully or reported to the homeowners' association for official rebuke.

Of course, in many countries, importing plants, plant parts, fruit, vegetables, flowers, *and seeds* is strictly forbidden. Contravening such phytosanitary regulations can make for a very expensive and time-consuming stopover at airport customs. The takeaway is that there are no doubt plenty of very attractive native plants for you to beautify your garden to your liking.

Salisbury, A. et al. 2015. EDITOR'S CHOICE: Enhancing gardens as habitats for flower-visiting aerial insects (pollinators): Should we plant native or exotic species? *Journal of Applied Ecology* 52(5): 1156–1164. https://doi.org/10.1111/1365-2664.12499

35. **It's just dirt, so we're coming in with heavy earth-moving machinery and remodeling it to create a more eye-pleasing landscape with gravel walkways, some park benches, and a ping-pong table or two.**
   - "Soil is life, treat it right!"
   - That would really soil your reputation.
   - Hey, most of the food you eat three times a day stems from the soil!
   - Beyond the important role of soil in providing a habitat for larger burrowing animals and "ecosystem engineers" such as earthworms, it is home to countless species from many taxonomic groups. They are often classified according to size (micro-, meso-, macro-, megafauna), and all play important roles and have specific functions in the overall ecosystem. Although many people are aware of how beneficial earthworms are and can recognize a few insects such as ants and beetles, only biologists probably recognize the many other "just-barely-visible" members of the soil community by name (enchytraeids, microarthropods such as mites and springtails, diplurans, or nematodes).

     Less well-known (and invisible to the naked eye) soil organisms, both to researchers and to the public, are microorganisms. Based on numbers, they far outweigh all other organisms in the soil: a handful of healthy soil can contain more microorganisms than there are people on Earth. The importance of their roles cannot be overstated. These microbes help form the soil structure, decompose plant litter and organic pollutants, cycle nutrients such as nitrogen and phosphorus, sustain soil fertility, and promote plant growth. Soils also play a role in climate change: they are the largest reservoir of carbon in the terrestrial biosphere or, from another perspective, contain twice as much carbon as all the vegetation and the atmosphere combined. Accordingly, our treatment of soils can determine whether they help to sequester carbon underground or join the other side and function as a source of greenhouse gases. Some soil microbes can also decompose microplastics (not to mention antibiotics), a hot topic these days. Their overall genetic biodiversity may be an important resource that we can use for specific tasks in the future. Of course, that biodiversity and those functions sink to nearly zero when covered by a thick layer of asphalt or cement. Moreover, sealing the soil or any natural landscape not only eradicates the fauna and flora but also destroys its ability to absorb water and replenish groundwater, directly promoting flooding. No need to mention that the micro-climate at such built-over sites will be altered indefinitely.

Coleman, D.C. et al. 2024. Chapter 5 – Soil fauna: Occurrence, biodiversity, and roles in ecosystem function. In: Paul. E.A., Frey, S.D. (eds.), *Soil Microbiology, Ecology and Biochemistry* (5th ed., pp. 131–159). Elsevier. https://doi.org/10.1016/B978-0-12-822941-5.00005-3

https://agriculture.basf.com/global/en/media/public-government-affairs/did-you-know/stories/handful-of-healthy-soil

Lal, R. et al. 2021. The role of soil in regulation of climate. *Philosophical Transactions of the Royal Society B* 37620210084. http://doi.org/10.1098/rstb.2021.0084

36. **These fields have only grass and weeds growing on them, so nothing much is lost by building our hotel complex/industrial park/parking lot here.**
    - Yet another habitat is to be paved over and permanently destroyed?
    - That sounds like "let's destroy the micro-climate first and see how that affects the global climate."
    - "Fields" are important ecosystems in their own right (grasslands, savannahs, prairies), supporting entire food chains (think buffalo in the USA). They may also represent the crucial initial growth phase in a plant succession that can ultimately lead to a climax stage, which can

potentially be similar to or indistinguishable from the original primary forest. Fields in the widest sense also provide a home for many burrowing species (again, think prairie dogs in the USA). Beyond such charismatic species, the top layers of soil typically contain an incredible abundance and density of fauna and flora (not to mention fungi). More than 95% of soil functions, however, are delivered by microorganisms (see also entry Nr. 35).

And speaking of grass, few people fully appreciate that the grass family (Poaceae) is one of the largest plant families on Earth (and *in* the "earth," if we want to nit-pick linguistically). Nearly 800 genera have been described, with potentially somewhere in the vicinity of 13,000 species. Some species included in this taxon are key agricultural plants such as wheat, rice, maize, and oats, making grasses the most economically important plant family. Bamboo also belongs here. Accordingly, grasses constitute over 40% of the land area on our planet. And speaking of "weeds," that derogatory category is an entirely artificial grouping. It encompasses a wide range of plants that, for one reason or another, are considered undesirable in a particular setting. A case in point is home gardens, where many companies make lucrative profits by providing chemicals to eradicate them. In the natural environment, however, such so-called weed plants are often an integral part of the ecosystem, for example, improving soil quality, sheltering organisms, repelling pests, or attracting beneficial insects. Some have medicinal properties, others are herbs or a meal in themselves. Once again, converting fields into urban landscapes or sealing them with buildings, parking lots, and roads destroys their ability to absorb water and replenish groundwater, directly promoting flooding. By the way, the ecological value of agricultural fields can be significantly increased by planting field margins, for example, with hedgerows.

https://en.wikipedia.org/wiki/Poaceae

Montgomery, I. et al. 2020. Hedgerows as ecosystems: Service delivery, management, and restoration. *Annual Review of Ecology, Evolution, and Systematics* 51: 81–102. https://doi.org/10.1146/annurev-ecolsys-012120-100346

37. **Removing all these big trees gives other trees a better chance to grow.**
    - So, you think you can design an ecosystem better than nature can?
    - Maybe, but someone else is profiting from that timber a whole lot more (and it probably isn't you).
    - One method of timber harvesting is termed selective logging, in which the largest or most valuable trees are taken. This tends to be a short-term

economic gain rather than a solid forest-management strategy. The method can leave poor material behind for forest regeneration. It also means a loss of seed sources because the biggest trees tend to produce the most seeds, and the same holds true when the "best" species are cut before they produce seeds or saplings. Although initially providing the highest profit, that one commercial advantage may only be temporary and fail to provide continuous profits in the long run. Forest sustainability is compromised because some tree species regenerate poorly after selective cutting. Finally, the heavy logging machinery disturbs the forest floor and topsoil. On the other hand, selectively logging certain species (e.g., Norway spruce) while retaining species with greater conservation value (e.g., oak) could be a strategy to support recovering temperate deciduous forests, one of the most degraded biomes in Europe and the world. Of course, certain alternative logging systems, such as clearcutting, also have their severe ecological drawbacks. The choice of method (and targeted trees) is difficult and should be made based on state-of-the-art forest ecology knowledge rather than profit maximization (see also entry Nr. 38 regarding old trees and deadwood as biodiversity promoters).

https://eos.com/blog/selective-logging/

Nordén, B. et al. 2019. The economy of selective cutting in recent mixed stands during restoration of temperate deciduous forest. *Scandinavian Journal of Forest Research* 34(8): 709–717. https://doi.org/10.1080/02827581.2019.1679876

38. **Removing these dead and decomposing trees opens space for other more attractive plants to grow.**
    - So you want natural forests to be manicured like suburban gardens?
    - Don't you think it's a bit presumptuous to intervene in the million-year-old cycle of life and death?
    - Dead and dying trees fulfil key roles in forests. One might distinguish three phases in a tree's life: growing, dead and erect, and lying on the forest floor. Together, they may last centuries. And each phase is essential for maintaining the forest ecosystem's biodiversity. In particular, dying and dead trees provide a habitat for a wide range of animals from insects to nesting birds and mammals. Such trees are typically the only ones providing a food source for woodpeckers, for example, and the tree-hollows that provide nests for birds and mammals are specific to older, larger trees. Deadwood also plays a key role in forest soil ecology (e.g., via wood-inhabiting fungi and arthropods). Larger-scale removals of deadwood for firewood, charcoal production, or biomass-based fuels

therefore reduce biodiversity and hamper forest health and regeneration. The need to prevent deadwood removal has been recognized as being crucial in tropical, temperate, and boreal forests. Management strategies here can include banning logging; protecting large, old trees; and actively enriching deadwood during harvesting by leaving tree parts behind.

Thorn, S. et al. 2020. The living dead: Acknowledging life after tree death to stop forest degradation. *Frontiers in Ecology and the Environment* 18(9): 505–512. https://doi.org/10.1002/fee.2252

39. **Those trees may be nice, but we really need the parking spaces and lighting.**
    - "They paved paradise and put up a parking lot." (Joni Mitchell, Big Yellow Taxi)
    - Define "need" (and who stands to make a profit – it probably isn't you).
    - Paving over any habitat decreases its former biodiversity to zero. Would you rather enjoy a habitat with plants and animals or a habitat for cars? As much as 65% of the total impervious ("waterproof") cover over America's landscape consists of streets, parking lots, and driveways. An estimated 5% of all urban land in the USA is covered by parking lots, more land than the states of Rhode Island and Delaware combined (eight spaces for every car in the country). Moreover, surfaces covered by cement, asphalt, or buildings are "heat islands"; prevent rainwater from being taken up by the soil and plants; and also hinder groundwater replenishment. The building and paving materials themselves can also release pollutants into stormwater runoff (as does the treated wood used for utility poles, railroad ties, etc.). Importantly, this form of soil compaction also increases uncontrolled runoff after major storms, causing flooding and landslides that are increasingly threatening human communities.

Bolin, C.A., Smith, S.T. 2011. Life cycle assessment of pentachlorophenol-treated wooden utility poles with comparisons to steel and concrete utility poles. *Renewable and Sustainable Energy Reviews* 15(5): 2475–2486. https://doi.org/10.1016/j.rser.2011.01.019

Frazer, L. 2005. Paving Paradise: The peril of impervious surfaces. *Environmental Health Perspectives* 113(7): A456–A462. https://doi.org/10.1289/ehp.113-a456

https://www.archdaily.com/976069/when-5-percent-of-the-united-states-is-covered-by-parking-lots-how-do-we-redesign-our-cities

40. **Why can't I walk on it – it's only moss/sand/corals! (Figure 2).**

Figure 2 "Leave only your footprints" may sound like a universally valid admonition in any nature conservation matter. Unfortunately, footprints by the thousands can damage many a sensitive ecosystem, including sand dunes and their vegetation, in a phenomenon known as trampling.

- Would you let groups of people walk on your fingers, which are no doubt more robust than moss?
- Don't tread on me (The Gadsden flag, American Revolution).
- There's lots of talk about the human footprint these days. A Human Footprint Index (HFI) has been developed to better quantify the total energy, resources, and products that a human utilizes during a lifetime. The issue here, however, is the physical imprint we leave on the ground we walk upon. Yes, even this can be a nature conservation matter. Excessive walking on sensitive substrates or vegetation is a phenomenon officially called trampling. This has been identified as a significant threat to habitats as diverse as Arctic shores, marine intertidal communities, sand dunes, and coral reefs (see Entry 86). The general impacts are loss of organic matter, damage to plants, change in vegetation species composition, reduction of soil porosity, decreased water permeability, increased runoff, and accelerated erosion. On rocky shores, for example, trampling can impact algae, barnacles, and mussels. On dunes, the management measures to prevent damage to vegetation and soil-dwelling insects, as well as to reduce soil compaction and erosion, include fencing, elevated walkways to enable access

to the water, and banning all-terrain vehicles. The recovery of trampled habitats, for example, in arctic regions, may take decades (see also vehicle disturbance, entry Nr. 57).

Brosnan, D.M., Crumrine, L.L. 1994. Effects of human trampling on marine rocky shore communities. *Journal of Experimental Marine Biology and Ecology* 177(1): 79–97. https://doi.org/10.1016/0022-0981(94)90145-7

Farris, E., Pisanu, S. et al. 2013. Human trampling effects on Mediterranean coastal dune plants. *Plant Biosystems – An International Journal Dealing with All Aspects of Plant Biology* 147(4): 1043–1051. https://doi.org/10.1080/11263504.2013.861540

41. **A well-kept plantation is just as ecologically valuable as the rainforest.**
   - If it's valuable because it's also green, then let's improve the environment by painting all road surfaces green.
   - Be selfish – are you the one enjoying the profits from the plantation?
   - While a plantation may be just as green from above, it no doubt contains only a very tiny percentage of the species composition of the forest it replaced. In fact, that tiny percentage may be composed entirely of the single species being cultivated, whether it be oil palm, rubber or banana trees, timber or pineapples. That's why it's referred to as a monoculture, which is designed in part to reduce resulting losses by competing species (and boost profits). What are other potential impacts of (almost all) terrestrial monocultures? Susceptibility to disease, invasive potential in adjoining areas, introduction of exotic or altered genotypes, altered habitat structure (e.g., canopy cover and shading), loss of ecological niches for wildlife, attraction of pests, pesticide and herbicide use, altered soil quality and soil communities, water (irrigation) demands, higher risks of adverse effects of storms and fire, vehicle use, as well as lower resistance to the biotic and abiotic disturbances aggravated by changing climates. Importantly, plantations are designed for regular harvests and no natural recovery is foreseen. This is also true of the increasingly touted "plantation forestry" (mixed-species plantations or rainforest plots for timber). Finally, the social impacts of large-scale plantations involve ownership changes from local communities to large private companies and a loss of traditional goods and cultures, customary rights, and livelihoods due to resettlement and unequal distribution of resources.

Cornwell, E. 2014. Effects of different agricultural systems on soil quality in Northern Limón province, Costa Rica. *Revista de Biologia Tropical* 62(3): 887–897. https://www.scielo.sa.cr/scielo.php?script=sci_arttext&pid=S0034-77442014000300005

Corsa Lok Ching Liu et al. 2018. Mixed-species versus monocultures in plantation forestry: Development, benefits, ecosystem services and perspectives for the future. *Global Ecology and Conservation* 15. https://doi.org/10.1016/j.gecco.2018.e00419

Kanowski, J. et al. 2005. Consequences of broadscale timber plantations for biodiversity in cleared rainforest landscapes of tropical and subtropical Australia. *Forest Ecology and Management* 208(1–3): 359–372. https://doi.org/10.1016/j.foreco.2005.01.018

42. **We don't really need all these native fruits and vegetables because other, better crops grow just as well here.**
    - If "variety is the spice of life," then you've just opted for a truly bland existence.
    - Other and better means "more profitable."
    - Potatoes forever!
    - That's an important issue because addressing food security and food shortages is an important side of the coin. That may involve growing crops that originally stem from other parts of the world or importing the products from distant countries. The other side of the coin is the cost of transporting such goods around the world. The most sustainable strategy is clearly to use local plants rather than transporting the products from afar. Humans cultivate only about 150 of an estimated 30,000 edible plant species worldwide; only 30 species make up most of our plant-based diets. Focusing only on those few species helps explain why 1 billion people go hungry, 2 billion are obese or overweight, and a crop failure can spell disaster. Commercially exploiting local wild species as crops is one practice that can help counter some of the above problems. The added advantages include:
      - improved food security,
      - strongly reduced transportation costs (reducing our carbon footprint),
      - a diversification of local crops,
      - maintaining the local plant-soil-microbe interactions (reducing soil degradation),
      - cashing in on the "novel food" craze,
      - reducing the impact of the so-called Western diet,
      - a contribution to diversifying global food production, and
      - promoting crops better adapted to local conditions and ultimately to the full range of environmental conditions that we inhabit.

Shelef, O. et al. 2017. The value of native plants and local production in an era of global agriculture. *Frontiers in Plant Science* 8: 1–15. https://doi.org/10.3389/fpls.2017.02069

43. **We don't really need these original native forms because we have already designed and bred much better varieties.**
- You mean you want to exchange the native forms, tried and tested over millennia, with a form created in the lab by ambitious young graduate students?
- Genetic mutants forever!
- Virtually all modern crop species have been selectively bred to yield the varieties we find on the market today. Importantly, more recent, genetically modified forms are also all based on native local forms somewhere or on their subsequently bred varieties. In many cases, however, researchers need to return to wild forms to study the genes that enabled their original success (disease resistance, nutritional value, withstanding changing or harsh environments). This has made finding and preserving local forms (and their seeds) a crucial endeavor. The resulting genebanks secure the genetic resources that underpin the food supply for future generations. A case in point is the potato, for which the International Potato Center (CIP) genebank maintains 1567 wild potato accessions, including 112 different *Solanum* species collected in 12 countries in the Americas. No single wild potato species extends throughout the total geographic area or altitudinal range (sea level to 4500 m). More broadly, we need to seek the best trade-off between (a) maintaining crop diversity at home by retaining native species, (b) improving crop diversity at home by introducing new species, and (c) investing more resources into increasing the productivity of existing crops (see also entry Nr. 42).

https://cipotato.org/genebankcip/

Huaman, Z. et al. 1997. *The Potato. Biodiversity in Trust Conservation and Use of Plant Genetic Resources in CGIAR Centres*. Cambridge University Press. pp. 21–28. https://doi.org/10.1017/CBO9780511470851.003

5
Freshwater

44. **These cormorants are eating all the fish and leaving nothing for us. Kill 'em all!**
    - Actually, we're stealing the fish from the cormorants.
    - You mean we've opened up what amounts to a restaurant for the birds and are surprised when they come?
    - This is a complex story. On the one hand, the fish populations in many lakes are the result of stocking fish, often for fishers. This is an artificial situation that has altered both the fish species composition and abundance. The same holds true for aquaculture ponds stocked with only a single species of fish. This creates an unnatural situation that both attracts cormorants and extends their presence on site. On the other hand, cormorants have also been documented to avoid larger predatory fishes preferred by commercial fishers and anglers, and preferentially preyed on smaller members of the carp family. This can even have a favorable effect regarding biological management plans, for example, by improving water transparency because the birds reduce the density of the bottom-feeding and sediment-suspending carp. Improved transparency, in turn, could negatively affect the cormorants by causing the fish to retreat to deeper water layers. In another case, however, cormorant predation was thought to be responsible for the major decline of brown trout in Danish streams. Management strategies for conflicts between cormorants and fisheries can be developed on the local or regional (e.g., EU) level. In a first step, ("serious") damage should be defined and quantified, at the ecosystem level, at the enterprise/income level, or regarding the amenity value. Potential actions include nest destruction, destroying or disturbing roost places, or killing breeding adults. The consensus, however, is that the large-scale killing of the birds is not an appropriate or promising measure.

DOI: 10.1201/9781003757474-5

Dirksen, S. et al. 1995. Cormorants *Phalacrocorax carbo sinensis* in shallow eutrophic freshwater lakes: prey choice and fish consumption in the non-breeding period and effects of large-scale fish removal. *Ardea* 83: 167–184.

Jepsen, N. et al. 2018. Change of foraging behavior of cormorants and the effect on river fish. *Hydrobiologia* 820: 189–199. https://doi.org/10.1007/s10750-018-3656-2

Marzano, M. et al. 2013. Managing European cormorant-fisheries conflicts: Problems, practicalities and policy. *Fisheries Management and Ecology* 20(5): 401–413. https://doi.org/10.1111/fme.12025

45. **There are just not enough fish here, so you should stock these waters.**

- Define enough, and enough of what exactly (and for whom?).
- So you want us to put fish in just so you can kill and eat them a few days or weeks later?
- In a natural or semi-natural aquatic system, the populations of the potentially very many fish species will have reached some sort of more or less balanced state. Any human-induced changes, such as stocking, are bound to upend that balance. Stocking typically involves larger, predatory fish (often only a single species such as trout) for recreational purposes. This can have multiple repercussions:
  - other top predators can be replaced;
  - native predators may be replaced (if an "exotic" species is stocked),
  - food web structure may change;
  - species on the lower rungs of the food chain face increased pressure (e.g., shifts in composition and abundances);
  - plankton communities may shift;
  - nutrient cycles can be altered; and
  - links between the aquatic and adjoining terrestrial system may be altered (e.g., reduced amphibians).

In one of the most notorious cases, introducing the large Nile perch into African lakes caused the extinction of numerous smaller endemic fish species. Stocking trout into small, historically fishless mountain lakes eliminated large invertebrate predators and amphibian species. Of course, stocked fish are often released only shortly prior to the fishing season, so that after a few months, the original situation may once again have been reached due to fishers. Then, again, the problem might be perpetuated by regular, annual stocking programs. Here, economic considerations (fishing licences, tourism) typically overshadow conservation-related considerations. In some cases, of course, stocking may be helpful in conserving or re-introducing a threatened species.

Eby, L.A. et al. 2006. Effects of stocking-up freshwater food webs. *Trends in Ecology and Evolution* 21(10): 576–584. https://doi.org/10.1016/j.tree.2006.06.016

46. **There were actually more and fatter fish here when the food-processing/sewage plant was still discharging directly into the water.**
    - That's because all the wastes directly or indirectly landed up in the fish themselves.
    - Hope you didn't eat any of those fattened fish.
    - Enriching a water body with nutrients from farmland fertilizer runoff, livestock wastes, sewage, food-processing plants, or the like is a form of pollution termed (cultural) eutrophication. And it's true that a little bit of added nutrients can promote plant growth on land and in the water. In freshwater that would be algae and phytoplankton. The increased phytoplankton may promote zooplankton growth, jump-starting the food chain all the way up to the fish. But, as in most things related to toxicity, it's the dose that makes the poison. In one scenario with larger inputs, the algal growth may become excessive, and the water turns greenish or some other color reflecting the pigments in the algae (plankton blooms). Turbidity increases, bacteria take over to decompose the extra organic matter, oxygen is used up, biodiversity decreases, and the system becomes unstable. The ultimate symptom of eutrophication is oxygen depletion (hypoxia, anoxia), which triggers mass mortalities of aquatic organisms. So, too many nutrients can ultimately mean ecosystem collapse (referred to as "dead zones" on the seafloor in the marine environment). A typical symptom of artificially "fattening" fishes, even with "good" food as in aqua- or mariculture, is the destruction of the bottom communities below the facilities. It's hard to imagine that discharging or adding anything to a natural water body could ever yield an "improvement."

Qin, B. et al. 2013. Lake eutrophication and its ecosystem response. *Chinese Science Bulletin* 58(9): 961–970. https://link.springer.com/article/10.1007/s11434-012-5560-x

47. **OK, this fish may be an "alien," but it grows bigger than anything else here and makes good eating, so it would really be great if you would release them everywhere.**
    - You don't want aliens in your spaceship, and you don't want them in your lake.
    - The practice of bringing home and setting free seemingly attractive species from faraway destinations has a long history, beginning in the early days of seafarers. Unfortunately, every introduced alien species

has almost always had more unexpected and negative consequences than positive effects. And equally unfortunately, we seem not to have learned our lesson, even today. Beyond the many unintentional introductions, ongoing purposeful introductions are typically motivated by thoughtlessness or short-sighted selfish or commercial interests rather than overall ecological insights. This issue is compounded today by climate warming, which abets the survival and range extensions of many invasive species. Introducing specimens that are genetically modified to fulfil some human desire can add insult to injury. That bigger alien fish, for example, has probably gotten bigger at the expense of native competitors and the hapless prey organisms that normally inhabited the waterbody. The result can range from the elimination of the natural fish top predator to the displacement of amphibians to the destabilization of the entire aquatic ecosystem, not to mention the introduction of new diseases and parasites. In Europe, the introduction of the rainbow trout is one of the better-known cases.

Josefsson, M., Andersson, B. 2001. The environmental consequences of alien species in the Swedish Lakes Mälaren, Hjälmaren, Vänern and Vättern. *AMBIO: A Journal of the Human Environment* 30(8): 514–521. https://doi.org/10.1579/0044-7447-30.8.514

Stanković, D. et al. 2015. Rainbow trout in Europe: Introduction, naturalization, and impacts. *Reviews in Fisheries Science & Aquaculture* 23(1): 39–71. https://doi.org/10.1080/23308249.2015.1024825

48. **What's the purpose of making this beautiful artificial lake if we can't all drive our motorboats on it?**
   - Even an artificial lake can be a biodiversity hotspot.
   - You mean speedboats criss-crossing the lake all day will improve your beautiful nature experience?
   - As opposed to running waters where the water at every site is constantly being renewed/replaced, the exchange rate for the standing waters of ponds and lakes is very slow. This means that pollutants tend to be quickly washed away and diluted in streams and rivers but often remain trapped over the long term in lakes. This is also valid for the oil and gasoline that all motors – including the in- and outboard motors of pleasure craft – inevitably lose. More importantly, even artificial lakes will attract and be home to a wide range of aquatic organisms, including fish and waterbirds, both of which can be disturbed by boaters. The noise of motorboats, for example, is a documented stress factor for fish. Moreover, the waves (wake) caused by fast-moving boats can impact precious shoreline vegetation and promote erosion. For this and other reasons (e.g., the spread of invasive species such as clams attached to

the hulls of boats brought in from elsewhere), motorboats and boat races are being banned in some lakes. Finally, if you like to swim, you should know that being hit by a motorboat makes an ugly mess.

Amoser, S. et al. 2004. Noise emission during the first powerboat race in an Alpine lake and potential impact on fish communities. *The Journal of the Acoustic Society of America* 116: 3789–3797. https://doi.org/10.1121/1.1808219

Jacobsen, L. et al. 2014. Effect of boat noise and angling on lake fish behaviour. *Journal of Fish Biology* 84: 1768–1780. https://doi.org/10.1111/jfb.12395

49. **This dam is a blessing for the species living in the river. It can guarantee a steady downstream flow year-round.**
    - Constant flow was never a feature of any natural river.
    - Admit it, you need the electricity to run your electric toothbrush.
    - Dams are major interventions in rivers or streams. They interrupt the flow, can alter water temperature along with water quality and sediment regimes, and impact the connectivity of the water body with the surrounding landscape. They also pose a barrier to fish migrations, which can contribute to the decline or even extirpation of species that depend on longitudinal movements along the river continuum. So-called fish ladders, if at all feasible for the dam type and size, can never recreate the original conditions for all migratory species. A case in point is sturgeons. In the last half-century, their populations have dropped 94%, with 3 of the 27 species feared to be extinct. We have destroyed almost every river habitat they occupy. They are all migratory, and dams pose a key threat (not to mention pollution and fishing – yes, they are some of the most valuable animals on Earth, think caviar . . .). Although the pros and cons of dams can be nit-picked and certain folks do benefit, some river dams and the associated water diversions for other purposes have meant that water no longer even reaches the sea (Colorado River, USA). In other cases, the local marine fisheries have been severely impacted. The classic example is the Aswan Dam in Egypt (yes, you will be able to find a counter-expertise if you look hard enough . . .). Of course, dam breaks (due to structural damage, floods, wars, and other causes) have and will continue to pose a major threat to downstream communities, including human settlements.

On a more philosophical note, the growing "rights of nature" movement calls for recognizing the inherent, inalienable rights of ecosystems and natural communities to exist and flourish. Rivers have become a particular focus, whereby some countries have already revised their constitutions to recognize the rights of nature (Ecuador). Others recognized a river as a "spiritual and physical entity" (New Zealand) or granted legal rights to a lagoon (Spain). (See also entry Nr. 58 for ocean rights.) This notion doesn't seem all too far-fetched, considering that corporations, for example, are non-human entities with a wide range of rights (including that of a fair trial).

Dams, fish and fisheries: Opportunities, challenges and conflict resolution. 2001. FAO Fisheries Technical Paper 419. 167 pp. https://books.google.at/books?hl=de&lr=&id=JpIO_ijV1WcC&oi=fnd&pg=PA45&dq=dams+problems&ots=kLhu1qA59R&sig=TkyNBZXmccQtvOD3zHElamgt9T8#v=onepage&q=dams%20problems&f=false

Macfarlane, R. 2025. Does a river have legal rights? *The New York Times International Edition, Opinion*. May 31–June 1, 2025. p. 11.

Richter, B.D. et al. 2024. New water accounting reveals why the Colorado River no longer reaches the sea. *Communications Earth & Environment* 5: 134. https://doi.org/10.1038/s43247-024-01291-0

50. **If the alligators would stay in their swamps, everything would be OK, but now they're walking all over our golf courses and basking in people's swimming pools! (Figure 3).**

Figure 3 If you drain the swamps, where else are the alligators supposed to go? And shouting "Fore" very loud won't convince the alligator to move out of the line of your golf shot. Don't forget, they can probably outrun you over short distances (like between you and the golf cart). (Image from Shutterstock).

- Golf course "water traps" take on a whole new meaning!
- Hey, that might shorten your retirement time at the golf course community and save your family lots of money.
- "A few alligators are naturally of the vicious type and inclined to resent it when you prod them with a stick. You can find out which ones these are by prodding them." (Will Cuppy)
- As alligators are not known to make long-distance migrations across land, this is more often than not a simple case of having built a golf course or human settlement on or too close to their natural habitat. In some situations and for some organisms (birds, turtles, insects), golf courses can actually increase the "ecological value" (e.g., animal species diversity and richness) relative to surrounding lands. In the case of alligators, such land-use changes bring the reptiles and humans closer together, potentially posing a threat to the latter (but, of course, also vice versa). If it's your pool, who do you call? No, not "ghost busters." Your area will no doubt have some kind of animal control agency, and in Florida, for example, there's even a free Nuisance Alligator Hotline.

From the alligators' perspective, the land modified by encroaching humans can strongly alter their feeding patterns and prey species, thus affecting their body condition, their exposure to chemicals, and their overall role within the ecological community. Large-bodied predators can be instrumental in structuring ecological communities, and their behaviors can be dramatically impacted by land-use change. Of course, the situation is quite similar to that for many large terrestrial animals (such as large herbivores: think elephants) whose altered habitats have promoted human–wildlife conflicts. The solutions can range from better fencing to regular removals (culling and/or relocations) to improved tolerance (conservation education programs).

If you are primarily concerned about your health, then the greatest threat may come from the herbicides, pesticides, and insecticides used on golf courses. A new study shows that people living within three miles of such courses faced a two- to three-fold increased risk of contracting Parkinson's.

Krzyzanowski, B. et al. 2025. Proximity to golf courses and risk of Parkinson disease. *JAMA Network Open* 8(5): e259198. https://doi.org/10.1001/jamanetworkopen.2025.9198

Rosenblatt, A.E. et al. 2023. Golf course living leads to a diet shift for American alligators. *Ecology and Evolution* 13(9): e10495. https://onlinelibrary.wiley.com/doi/full/10.1002/ece3.10495

Skupien, G.M. et al. 2016. Teaching tolerance? Effects of conservation education programs on wildlife acceptance capacity for the American Alligator.

Human Dimensions of Wildlife 21(3): 264–279. https://doi.org/10.1080/10871209.2016.1147624

51. **I'd like to buy one of those baby alligators.**
    - If you want something exotic, buy a perfume with a seductive name or a fruit you've never heard of before.
    - Well, you should know that an alligator – and all crocodilians – are "exotic" pets and top predators to boot. That means you might lose your nose when you smooch with them like you would with your dog or cat (whose predatory instincts are not necessarily activated by your nose – oops, sorry, yes, there always will be exceptions). Alligator brains are smallish, and some might say relatively "primitive," and their behavioral repertoire (at least as far as "cuteness" is concerned) is limited: they won't wag their tails when they see you. You'll never know exactly what they are thinking and whether they might actually be sizing you or parts of you up as a meal. Importantly, "exotic" can mean that it's illegal to keep such an animal in your area or that you'll need a license. You should also know that many alligator "farms" that produce cute baby alligators utilize wild-sourced eggs collected from wetland habitats. This means the advertisement assurances that such facilities are working hard to contribute to nature conservation or species protection are dishonest. Importantly, most of these animals grow to menacing sizes. At some point, you'll no doubt have to get rid of your pet, and there are very few organizations that will take them off your hands. Thankfully, the days of buying stuffed real baby alligators in curio shops are also (almost) gone. And if you believe the promotion texts stating that buying a larger "mounted" alligator head really supports species and swampland conservation and "invokes nature's inspiration in your home, business or office," then this whole book is not for you. If you truly must have a reptile, consider a small lizard or a turtle. And if it's for the kids, they can cuddle up much better to a stuffed, fluffy toy alligator.

Nickum, J. et al. 2017. Alligator (*Alligator mississippiensis*) aquaculture in the United States. *Reviews in Fisheries Science & Aquaculture* 26(1): 86–98. https://doi.org/10.1080/23308249.2017.1355350

52. **There is no water shortage – the lakes are all full!**
    - If you wait a bit longer, you can take a mud bath in the remaining puddles.
    - Yes, that may theoretically be true because we do live on a blue (= water) planet. Unfortunately for us, only 2.5% of that is freshwater, and only 0.03% is contained in lakes and even less in rivers (the remainder is

held in ice and groundwater). That tiny amount of surface water is, of course, highly susceptible to losses, and the declining water content of many water bodies, especially shallow lakes and ponds, has been recognized as a serious issue. This is exacerbated by increasing water removals for the many needs we have (drinking water, agriculture, golf courses, car washes, toilet flushing). Climate-change-related pressures will also increase. Accordingly, lake monitoring efforts show distinct water level drops and increasing water temperatures around the world. Even whole lakes (so-called terminal lakes with no outlets) and major inland seas have dried out (Aral Sea). Others are shrinking fast (e.g., Lake Poopó, Bolivia; Lake Chad, Africa; Lake Urmia, Iran; Great Salt Lake, USA). Equally damaging, however, are excessive fluctuations in water level, whereby climate models predict an increased occurrence of extreme events such as flooding and extended droughts. Drying lakes are even associated with increasing air pollution (windblown lake-bed fine dust particles). Can we reduce our water consumption? You bet! Do you let the faucet run while brushing your teeth? Of course, even lakes that are "full" are not necessarily healthy. Size is no help ("the solution to pollution is dilution". . .): The world's deepest lake (Baikal) faces severe pollution problems, and even the Great Lakes (1/5 of the world's surface freshwater) need to be saved from pollution, over-fertilization, invasive species, and climate change.

Jones, B.A., Fleck, J. 2020. Shrinking lakes, air pollution, and human health: Evidence from California's Salton Sea. *Science of the Total Environment* 712: 136490. https://doi.org/10.1016/j.scitotenv.2019.136490

Saving the Great Lakes. 2020. National Geographic, 12.2020.

Zohary, T., Ostrovsky, I. 2011. Ecological impacts of excessive water level fluctuations in stratified freshwater lakes. *Inland Waters* 1(1): 47–59. https://doi.org/10.5268/IW-1.1.406

53. **These eels definitely reproduce in our freshwaters – after all, we always catch a few small individuals too. It's a myth that they spawn only in the sea.**
    - There's something fishy going on here.
    - Every big eel was once smaller (somewhere else).
    - Freshwater eel reproduction was long a mystery. It took centuries to determine that eels actually had reproductive organs (even famed academics such as Sigmund Freud – I'm not making this up – failed to find them). Scientists finally determined that these eels reproduce only once in their lifetime and that they needed to migrate from the adult freshwater habitat to the ocean to do so. Only decades later was the

Sargasso Sea in the Atlantic finally identified as one such spawning site. The freshly hatched larval eels then drift for several years with the ocean currents, covering thousands of kilometers to reach their freshwater habitats in Europe and the US East Coast, first as "glass eels," then as yellow eels or elvers. The latter show a strong tendency to swim against the current. They then mature for a decade or more before taking on the return trip to the sea. Finding small eels far upstream or in lakes does not mean they reproduced there. The intense market demand for young eels in Asia has made them the most valuable marine resource ever (in terms of "price per pound"), posing a severe risk to their overall populations (the IUCN lists the species as "critically endangered"). Prompted by the potential billion-dollar profits, several institutions in Europe have been competing for years to artificially "breed" the European eels. All the most modern techniques available have failed: the record for keeping glass eel "babies" alive is only a few hundred days, and their further development is still unachieved. Their long migration in different Atlantic currents under different light conditions, water temperatures, salt concentrations, and (largely unknown) food sources simply cannot be recreated. The eel is the only fish to have a separate EU regulation devoted to it (EC No 1100/2007). Accordingly, the capture of eels should be prohibited, the illegal export to Asia should be stopped, and everything should be done to keep our rivers and streams flowing as naturally as possible. Where they are dammed, fish ladders should be installed to help enable the eels (and some of the many other migrating fish species, e.g., salmon) to complete their life cycles. The take-home message? Recreating natural processes (ecosystem functions, ecosystem services) can be very expensive, if at all possible.

https://www.maine.gov/dmr/fisheries/commercial/fisheries-by-species/eels-and-elvers/the-maine-eel-and-elver-fisheries

Katsumi, A. et al. (eds.). 2003. *Eel Biology*. Springer Japan. 497 pp. https://doi.org/10.1007/978-4-431-65907-5

54. **These beavers are killing our trees, flooding adjoining property, and hindering my kayaking – we should shoot them all.**
    - Who chops down more trees, beavers or humans?
    - If you think being busy and industrious is praiseworthy, then "eager beaver" or "busy beaver" will no doubt be at the top of your list.
    - So, you'd rather have an arid habitat than a lush wetland?
    - "Beaver do better work than the (US) Corps of Engineers." (Mike Todd)

- Over the centuries, beavers were extirpated from most waterways in the world for their pelts. Today, they are making a comeback, and some consider them to be pests (even though their population in North America, for example, is only 10% of what it was originally). The latest take is that they are climate heroes! How's that? Well, they are perhaps one of the best-known examples of "ecosystem engineers," species whose activity forms their habitat. In the present case, beavers provide advantages to the environment and services to humans. These advantages include:
  - improved water quality (by slowing the water flow);
  - storing and more gradually releasing water after rainfall or snowmelt, preventing some streams from drying out;
  - reducing nutrient loads such as nitrates from fertilizers and toxic pollutants such as heavy metals (beaver ponds help filter them out and store them in the sediment);
  - increased resistance to wildfires and drought (their ponds and canals irrigate the surroundings, creating a lush landscape that provides a refuge for wildlife and livestock);
  - reduced flooding in beaver-created wetlands;
  - more carbon stored than in adjoining habitats; and
  - felled trees making way for new growth.

  Not to mention providing habitat for numerous aquatic and semi-aquatic species from insects to amphibians to birds, salmon, and moose. Sounds like great arguments for letting beavers restore your wetlands – for free. "Leave it to Beaver"! (popular American sitcom in the 1950s/1960s).

  Of course, in certain situations, beaver activity may run counter to other interests. This is no doubt more so the case in Europe, where cultural landscapes predominate and need to be maintained, often at great effort and cost. For example, the destruction by beavers of earthen walls in aquaculture fish ponds must be vigorously countered. Rarely, however, is killing them all the answer. Trapping and releasing them elsewhere is one strategy. Unwanted beaver-created flooding due to blocked pipes and culverts can typically be effectively countered with fencing. Preventing damage to particular trees can be simply achieved by enclosing the lowest part of their trunks with mesh wire.

Fairfax, E., Whittle, A. 2020. Smokey the beaver: Beaver-dammed riparian corridors stay green during wildfire throughout the western United States. *Ecological Applications* 30(8). https://doi.org/10.1002/eap.2225

Goldfarb, B. 2025. What we can learn from the genius of beavers. *National Geographic* 247(6): 78–99.

6
Marine

55. **They're only empty seashells – why shouldn't I take a bagful home with me?**
    - Each empty snail shell could be a hermit crab's new home.
    - You can hear the ocean without holding a shell to your ear.
    - Billions of people visit the ocean every year. If even only every second visitor collected a single shell (and some take home a bagful), that's an immeasurable pile removed from the ecosystem. The shells of all molluscs – bivalves (clams and mussels) and snails – play important roles in marine habitats. All are eventually broken down by predators, endolithic species (those that bore into harder material), or wave action, often becoming key components in the grains of "sand" making up beaches. In these days of major beach erosion, that's significant. In the case of snail shells, each is a potential home to a hermit crab. The internet is full of photos of hermit crabs occupying suboptimal man-made objects because the right shells are missing. Importantly, none of the countless, perfect, shiny shells for sale in curio and souvenir shops were dead when collected. They were all taken commercially as living molluscs and killed. Everyone should do their best to avoid purchasing glossy, intricate, or otherwise impressive shells – and even more so the figurines made of hundreds of glued-together molluscs (see also Figure 17). Some species are in danger of being overexploited, and some are on the endangered list and under protection. This means customs agents at the border may fine you stiffly for trying to export and/or import them. Other molluscs, such as the impressive giant clam, are also available in remarkably realistic synthetic materials. Making memories shouldn't mean taking something out of the ecosystem. Maybe a good close-up photo underwater or on the beach would make an equally satisfying memento of your vacation?

DOI: 10.1201/9781003757474-6

Chavan, A.S., Chondekar, R.P. 2023. Overexploitation of cowrie shells: Future threats to the marine ecosystem. *International Journal of Zoology Studies* 8(3): 8–10. www.zoologyjournals.com

Coppa, S. et al. 2023. What is hidden in the luggage? First assessments of illegal seashells gathering in Sardinia (Italy). *Conservation Science and Practice* 5(6): e12913. https://doi.org/10.1111/csp2.12913

56. **What the hell do we need sea turtles/sea cows for? What have they ever done for us?**
    - In that vein, what do we need Picasso, the Beatles, etc. for?
    - Well, tell me what you have done for humankind.
    - What would do this world more good: more sea turtles or more people?
    - Does our planet really need any people?
    - Sea turtles, like all organisms, play a role in their habitat. This role ranges from serving both as a predator and as a prey, to providing a habitat (their shells) for encrusting organisms, to fertilizing and shaping their nesting beaches ("ecosystem engineers"). Many such ecosystem functions are directly and indirectly relevant to us and can be termed ecosystem services. For example, some sea turtles (leatherbacks) feed almost entirely on jellyfish. Decimating these sea turtles can therefore contribute (in addition to pollution and warming waters) to more jellyfish, whose blooms can injure humans and impact tourist holiday destinations. This has clear negative effects on coastal and even national economies. Another sea turtle is among the few species that feed on sponges. In coral reefs, they keep space competitors such as sponges in check, helping to maintain a healthy reef ecosystem. A third sea turtle species grazes on older, often overgrown parts of seagrass leaves, helping to keep the meadows healthy. Major declines of these turtle species from historical levels, therefore, pose a threat to marine ecosystems of crucial ecological and commercial importance. Similar lines of argumentation can be used for sea cows (dugongs, manatees), which regulate the biomass, productivity, and composition of the algal and seagrass meadows they feed on.

Heithaus, M.R. 2013. Predators, prey, and the ecological roles of sea turtles. In: Wyneken, J., Lohmann, K.J., Musick, J.A. (eds.), *The Biology of Sea Turtles* (Vol. III, pp. 249–284). CRC Press.

Ruiz-Frau, A. 2023. Impacts of jellyfish presence on tourists' holiday destination choices and their willingness to pay for mitigation measures. *Journal of Environmental Planning and Management* 66(10): 2107–2125. https://doi.org/10.1080/09640568.2022.2061926

Wirsing, A.J. et al. 2022. Ecological roles and importance of sea cows (Order Sirenia): A review and prospectus. *Marine Ecology Progress Series* 689: 191–215. https://doi.org/10.3354/meps14031

57. Why can't I drive my car on the beach? This is a free country, isn't it?! (Figure 4).

Figure 4 Vehicles on beaches and dunes are predestined to scar the landscape and create havoc. If that doesn't sway you, the combination of sand and salt is the fastest route to prematurely age your car. Not to mention the incoming tide and costs of being towed out. (Image from Shutterstock).

- Nothing will devalue your beautiful car faster than sand and salt.
- What other ecosystem would you and others like to drive over?
- Fine, you'll be boosting the economy because you'll soon need to buy a new car.
- Beaches are living, moving ecosystems. They are also home to an enormous wealth of organisms ranging from the sand fauna in the moist beach stretches to seabirds and dune vegetation in the upper parts of the shore. Driving cars there, during either the day or night, is a major disturbance to wildlife and is strictly banned almost everywhere. On sea turtle nesting beaches, for example, vehicle traffic is a recognized threat. Beyond potentially running over nesting adult turtles or hatchlings on their way to the sea, the main issue is sand compaction over nests. This hinders the hatchlings from digging their way out to the

surface, potentially killing the entire clutch. Moreover, hatchlings can become misoriented in vehicle tracks long after the vehicles have passed by: they run parallel to the sea inside the tracks until they die of exhaustion or heat. On very long beaches, sea turtle researchers drive their quads outside the nesting zone itself (for example, directly along the water line) during their daily monitoring efforts. Elsewhere, access to remote beach sites with vehicles is sometimes permitted for licensed fishers. The motivations of other drivers may be alcohol- or testosterone-fueled or otherwise iffy. Clearly, the excuse for motoring along a beach to experience quiet and remote coastal landscapes is schizophrenic at best. Importantly, driving a vehicle across any natural landscape is bound to create damage.

Aguilera, M. et al. 2019. Assessing the effects of multiple off-road vehicle (ORVs) tyre ruts on seaward orientation of hatchling sea turtles: Implications for conservation. *Journal of Coastal Conservation* 23: 111–119. https://doi.org/10.1007/s11852-018-0641-x

Kevan, P.G. et al. 1995. Vehicle tracks on high Arctic tundra: Their effects on the soil, vegetation, and soil arthropods. *Journal of Applied Ecology* 32: 655–667. https://doi.org/10.2307/2404660

Petch, N. et al. 2018. Motivations and behavior of off-road drivers on sandy beaches. *Ocean & Coastal Management* 163: 82–91. https://doi.org/10.1016/j.ocecoaman.2018.05.021

58. **The oceans are huge – you can't tell me we are really degrading them!**
    - Name me one form of pollution that hasn't been reported in the oceans.
    - No ocean, no us. (Sylvia Earle)
    - The simple truth is that almost every type of pollution ends up being marine pollution (and the only thing that leaves the ocean is distilled water in the form of evaporation).
    - It's alarming that we are degrading a habitat in which we don't even live. In many contexts, any mention of the oceans or of "man and the sea" immediately sparks a discussion about marine pollution. The pollutants include crude oil, heavy metals, halogenated hydrocarbons, radioactivity, marine debris (plastics), and organic wastes (eutrophication). Importantly, pollution is only one side of the coin. The other is our many other uses of the oceans, most of which cause some kind of damage or disturbance. These include using them directly as a dump, criss-crossing them with shipping routes (noise, accidents, collisions with sea life), exploiting and over-exploiting marine resources (think oil rigs, fisheries, or potential future deep-sea mining), spoiling shorelines with barriers and high-rise apartments, and lastly, our many

recreational and sporting exploits. Almost no sea region is unimpacted. In fact, researchers are increasingly having difficulty finding and studying pristine environments to determine how marine ecosystems function naturally. Instead of conveying how marine ecosystems function, ever more research papers and university teaching courses today are devoted to explaining why natural ecosystems no longer function as they should and what actions must be taken to alleviate the damage. Entire high-ranking scientific journals are devoted to marine pollution (Marine Pollution Bulletin). The nature documentaries that dazzle us with their beauty are often misleading: the film-makers must often seek ever more remote corners of the world to present intact environments to viewers. This viewpoint is supported by today's vacationers, who are booking the most faraway, exotic (and expensive) destinations in the hopes of enjoying "pristine" nature.

The multifaceted decline in the state of the world's oceans is yet another argument in support of the rights of nature movement, in this case the attempts at recognizing the ocean as an independent legal entity (www.oceanrights.com). This would provide additional legal instruments to protect it for its own sake and thus advance an ethical, respectful, and reciprocal relationship with humanity (see also entry Nr. 49).

https://www.sciencedirect.com/journal/marine-pollution-bulletin

Stachowitsch, M. 2003. Viewpoint: Research on natural marine ecosystems: A lost era. *Marine Pollution Bulletin* 46: 801–805.

59. **Those damn sea turtles/dolphins/sharks are damaging our nets and stealing the fish: we have to kill them to protect our jobs.**
    - Do you think the oceans belong more to them or more to us?
    - Are they stealing our food or are we stealing their food?
    - This is a phenomenon referred to as depredation, which in the marine environment describes when larger predators – everything from sea lions to orcas – feed on the bait that has been laid out or on the fishery catches themselves. What most animals learn fast is where food is to be found, the easier the better. And bait dangling from a hook or the fish already hooked on longlines are effortless meals. Other predators learn to follow fishing vessels and consume the fish stunned, injured, or fleeing from heavy trawling nets. This poses certain risks to the predators themselves, which may become hooked, entangled in lines, and be killed by the fishers if not already drowned (marine mammals, sea turtles, or sea birds that breathe air and dive down to the snared fish). This, alas, is yet another human–wildlife conflict. One that also

poses a threat to fishers, who take a risk in trying to entangle still-living, larger predators (and who lose valuable time in untangling dead animals). In the fishing industry, money and jobs also play a major role. Economically, each hooked and unwanted animal means one less hook for the desired catch. The purported threat of job loss is one that resonates in every segment of the public, regardless of whether it is in a conservation context or in some other framework. This issue often requires detailed analysis to relativize – only rarely are livelihoods truly endangered by depredation. In many cases, the conflict can be de-escalated with modern (and often simple) management strategies and technologies. These can include installing turtle excluder devices (TEDs) in trawl nets or positioning longline hooks somewhat deeper to put them out of reach for most air-breathing predators.

Tixier, P. et al. 2021. When large marine predators feed on fisheries catches: Global patterns of the depredation conflict and directions for coexistence. *Fish and Fisheries* 22(1): 31–53. https://doi.org/10.1111/faf.12504

60. **We have to support ailing natural coral reefs by creating artificial reefs, and this oil platform/this decommissioned ship would make the perfect artificial reef (Figure 5).**

Figure 5 Most artificial reefs made of metal are deployed to get rid of large "retired" items on the cheap. The coral colonies growing on the hull of this decommissioned and purposefully sunk ship will all eventually perish in a rusty heap.

- It's made out of metal, right? And what happens to metal in saltwater?
- Heads up if such claims are made by the oil industry or ship owners.
- Artificial reefs have been a hot topic for some time. They are nonetheless quite controversial. First of all, many who wish to "sell" the idea of such reefs to the public or authorities are actually hoping to get rid of their "junk" on the cheap. Instead of dismantling an oil rig after its useful service life, they strive to leave large parts of it standing as an artificial reef. Instead of safely and properly scrapping an old ship, they want authorization to simply sink it somewhere. Others wish to dump anything from old subway cars to army tanks in the sea. Currently, the world's largest planned artificial reef is the ocean liner SS United States (measuring 300 meters and once the fastest ship in the world). It has been rusting away for 50 years and is apparently destined to be sunk off the US coast in the Gulf of Mexico. Another strategy is declaring such endeavors as art, for example creating an aquatic sculpture park made of freshly cast concrete objects (e.g., "cars") sunk off the coastline of Miami, Florida (Reefline). A classic case gone awry is using bundled and tethered old car tires (which no one has really found a good use for) as artificial reefs along the East Coast of the USA. That failed miserably because major storms soon tore them apart, and recovering the loose tires that were rolling over the seafloor and damaging the environment cost many millions of dollars.

Positive things can be said about some (non-toxic and preferably non-rusting) artificial structures if placed next to coral reefs to attract divers away from the living reef and thus reduce diver-related damage to the habitat (see entry Nr. 86). Another often posited argument is that they attract fish of interest to sport fishers. It is often unclear, however, whether they merely attract or concentrate fish from the surroundings, or whether they actually increase the relevant fish numbers and biomass. A more general positive aspect is that larger artificial reefs are a no-go area for commercial fisheries, whose nets can snag on and be destroyed by such structures. This can enable the immediately surrounding seafloor to recover from destructive fishery practices (trawling, dredging). Unfortunately, the ultimate fate of all metal-based reefs in the sea is a pile of rust (and a relatively short-lived life for any encrusting organism communities growing on them).

https://www.thereefline.org

Lima, J.S. et al. 2019. Overview and trends of ecological and socioeconomic research on artificial reefs. *Marine Environmental Research* 145: 81–96. https://doi.org/10.1016/j.marenvres.2019.01.010

Seaman, W. Jr. 2000. *Artificial Reef Evaluation. With Application to Natural Marine Habitats*. CRC Press. 246 pp.

61. **It would be a crime not to fully develop this beautiful reef site so that as many tourists as possible can visit it and learn to appreciate corals.**
   - Show me the environmental laws governing construction at that site.
   - Building something directly bordering a beach or within eyesight of a reef probably borders on illegality and is the bigger crime.
   - Show me one fully developed tourist site in which the adjoining reef has benefited.
   - Developing a site for tourism is, of course, a two-edged sword. On the one hand, bringing people closer to nature and nurturing their understanding of the natural environment is positive. On the other hand, any such project will be motivated by and firmly anchored in commercial interests, and the responsible businesspeople and backers will tend to put environmental protection and conservation relatively low on their list of priorities (despite fervent lip service to the contrary or blatant greenwashing). Nature will almost always take a backseat to human greed. Equally certain is that many future guests at the projected beachside hotel complex come for reasons other than helping to promote marine conservation. And each guest will contribute to the overall waste produced and to the freshwater demand. Most important from a reef protection standpoint, however, is that waves of tourists will change the environment. Indeed, virtually every individual diver and snorkeler will leave a negative "footprint" in the reef (see entry Nr. 85). How many people are too many? One answer is to think about the "carrying capacity" of a reef. This means how many dives should be permitted in a particular reef over the course of one year before the damage becomes unacceptable. The problem is that defining an acceptable level or index is a truly troublesome task. Some sites tally several hundred thousand dives per year – a recipe for disaster!

Fernández, P.R., Rivera, A. 2024. SCUBA diver behavior, diver carrying capacity and its implications for sustainable reef management. *Marine Policy* 165: 106199. https://doi.org/10.1016/j.marpol.2024.106199

Gil, M.A. et al. 2015. Rapid tourism growth and declining coral reefs in Akumal, Mexico. *Marine Biology* 162: 2225–2233. https://doi.org/10.1007/s00227-015-2748-z

Jameson, S. et al. 1999. A coral damage index and its application to diving sites in the Egyptian Red Sea. *Coral Reefs* 18: 333–339. https://doi.org/10.1007/s003380050208

Trave, C. et al. 2017. Are we killing them with kindness? Evaluation of sustainable marine wildlife tourism. *Biological Conservation* 209: 211–222. https://doi.org/10.1016/j.biocon.2017.02.020

62. **We want to improve the way the ocean functions by fertilizing it with trace metals like iron to promote productivity and combat climate warming**
(see entry Nr. 190).
63. **We have to dump our garbage at sea because there is no place on land/ our island is too small for burying or burning it.**
    - You mean out of sight, out of mind?
    - Yet another contribution to the 10 million tons of plastics that enter the oceans every year (that's 15,000 kilograms or 33,000 pounds per minute).
    - Disposing of garbage is a major problem facing humanity today. Some is burned ("thermal treatment"), the rest has to be dumped somewhere, and only relatively tiny amounts are truly "recycled." Authorized dumping can be expensive and requires space, and the garbage often contains toxic and other problematic materials. This might help explain why so much of it is being shipped around the world to less developed countries for what might be termed suboptimal disposal. The problem of space and the strong resistance to waste disposal sites close by (the NIMBY – not in my backyard – phenomenon) leads to casting an eye out to sea. In the past, ocean dumping was practiced on a grand scale for everything, including nuclear wastes (somewhere over 100,000 barrels full of radioactive wastes were dumped off the Pacific and Atlantic coasts of the USA and off Europe!) and munitions (chemical and conventional: millions of tons) after major wars. Sea charts today depict many of the latter shallow-water sites as posing a threat and being off limits to anchoring and fishing.

      Of course, many dumps on land and almost all those underwater have turned out to be major environmental mistakes that require rectifying. And this is the crux of the matter. Dumping on land typically involves a distinct, well-enclosed, and, if necessary, well-guarded area. The idea is to contain the wastes (and to be able to dig them up again if necessary). At sea, sites are never enclosed, and their position and bounds are not as distinct as on land. The idea here is often not to contain the waste, but rather to enable its gradual dilution (e.g., chemical wastes) until they reach "harmless" concentrations. In aquatic environments, a standard phrase of waste engineers has been "the solution to pollution is dilution" (see entry Nr. 151). In the present context, this means that land dumps can theoretically, albeit at great cost, be remediated or the material unearthed and moved elsewhere. This can never be a real option, either financially or technically, for waste at the bottom of the ocean.

Harrison, J. 2015. C. Ocean dumping. *Yearbook of International Environmental Law* 26: 196–204. https://doi.org/10.1093/yiel/yvw010

Ketchum, B.H. et al. (eds.). 2013. *Ocean Dumping of Industrial Wastes*. Plenum Press. 519 pp. https://books.google.at/books?hl=de&lr=&id=HuDgBwAAQBAJ&oi=fnd&pg=PA3&dq=ocean+dumping&ots=QlNxVjhZ4b&sig=5v73JXGNhJsVXDc11CHfwU3ArhQ#v=onepage&q=ocean%20dumping&f=false

Parmentier, R. 2014. Chapter: Radioactive waste dumping at sea. In: Sands, P. (ed.), *Greening International Law* (pp. 140–158). Routledge.

64. **These mangroves are blocking our access to the shore and view of the sea and should be removed** (see also entry Nr. 65).
   - Using that argumentation, all the high-rise buildings lining the coast should probably also be torn down?
   - Mangroves play key roles along many tropical coasts but have been significantly decimated. They may also represent almost the only forests and tree species along arid coastlines. A key driving factor for this deforestation has been the expansion of agriculture and aquaculture over the last 50 years, for example, to create space for mariculture projects such as shrimp farms. The decline is from an estimated 225,000 km^2 in the 1970s to about 137,000 km^2 in 2014. Although most of the losses occurred pre-2000, more recent satellite imagery showed that an estimated 62% of global losses between 2000 and 2016 resulted from land-use change, primarily through conversion to aquaculture and agriculture. Up to 80% of these human-driven losses occurred within six Southeast Asian nations, reflecting the regionality of the aquaculture boom (mostly for export) and the economic driving forces. Importantly, many shrimp ponds may last only a few years before having to be abandoned for numerous reasons, such as toxin or disease build-up. Accordingly, beyond their role in ecosystem function (see entry Nr. 24), the ecosystem services that mangroves provide, and thus their value to humans, exceed any commercial aquaculture profits by far. Although shrimp farming creates enormous private profits, it is less economically viable once the externalities generated by mangrove destruction and water pollution are included. Those externalities include their value as natural barriers protecting coastal communities against cyclones, hurricanes, tsunamis, and waves, which is estimated to encompass $82 billion of annual value globally. On a per hectare

basis, that value to the local community can be up to $30,000 (e.g., in southern Thailand, Indonesia), whereby the additional services include water purification, flood protection, erosion control, sediment stabilization, and prevention of salt interactions. Their role as effective carbon sinks (containing as much as four times more carbon when compared with other forest systems) is also increasingly being recognized as yet another marine bulwark against climate change. All this means that mangrove forests are worth more standing and healthy than when exploited or removed for other short-term gains. This recognition has led to significant mangrove transplantation efforts worldwide.

Bunting, P. et al. 2022. Global mangrove extent change 1996–2020: Global Mangrove watch version 3.0. *Remote Sensing* 14(15): 365. https://doi.org/10.3390/rs14153657

Goldberg, L. et al. 2020. Global declines in human-driven mangrove loss. *Global Change Biology* 26(10): 5844–5855.

Rizal, A. et al. 2018. Economic value estimation of mangrove ecosystems in Indonesia. *Biodiversity International Journal* 2(1): 98–100.

Sathirathai, S., Barbier, E.B. 2007. Valuing mangrove conservation in southern Thailand. *Contemporary Economic Policy* 19(2): 109–122. https://doi.org/10.1111/j.1465-7287.2001.tb00054.x

65. **We're not going to touch the reefs themselves; we only need to clear some of the mangroves for our project**
(see also entry Nr. 64).
- That's like cutting off your big toe but claiming you are pampering the rest of your foot.
- The coral reef ecosystem comprises much more than the corals themselves. The overall habitat mosaic typically includes the reef lagoon, sand bottoms, seagrass meadows, the open water, the adjoining mangrove forests, and the beach. All are interconnected with each other, and each plays a role in keeping the reef itself healthy. The roles of mangrove forests include:
  - a biodiversity hotspot in their own right (e.g., invertebrate colonies on their roots);
  - nursery grounds for juvenile fish and invertebrates from adjoining reefs;
  - capturing terrestrial runoff, that is, preventing sediments and many pollutants from reaching and damaging coral colonies;
  - producing leaf litter that supports the food web; and
  - providing nesting sites for birds and a habitat for insects, mammals, and reptiles.

Accordingly, coral reefs tend to be much better developed off mangrove forests than in unprotected adjoining areas. Clearing such forests poses a direct threat to the reefs, which are currently suffering multiple other threats. The combined effects of these threats (including bleaching due to climate warming) have some researchers predicting the demise of coral reefs across the globe. This calls for using a precautionary approach and eliminating any threats we can to the overall coral ecosystem. Refraining from cutting down mangrove forests is certainly one of the easiest and logical steps, as are mangrove protection projects and restoration programs, including planting.

Lovelock, C.E. et al. 2022. Tackling the mangrove restoration challenge. *PLOS Biology* 20(10): e3001836. https://doi.org/10.1371/journal.pbio.3001836

Nagelkerken, I. et al. 2008. The habitat function of mangroves for terrestrial and marine fauna: A review. *Aquatic Botany* 89(2): 155–185. https://doi.org/10.1016/j.aquabot.2007.12.007

66. Watching these dolphins here in the "marineland" pool helps gain public support for species protection (Figure 6).

Figure 6 What's more important, animal welfare or our entertainment? How much of an orca's natural behavior is being enabled here? Are visitors really going home with a new understanding and appreciation about species protection and nature conservation? (And yes, a little girl has been seated on the animal . . .).

- And what do you learn by watching a lion in a metal cage?
- Dolphins in various marine parks and public aquariums are always a major public attraction – and, like any public magnet, a money-maker for the facility. Nonetheless, this attraction has come under increasing criticism and many have closed. The critique centers around the practice of capturing animals from the wild (removing them from their family group) and around animal welfare considerations in captivity. The latter involve animal-based measures (behavior and health) and resource-based measures (e.g., enclosure space). Although animal welfare is more difficult to measure in marine than in terrestrial mammals, the general consensus is that many criteria for welfare are not and cannot be met for captive dolphins. The problems are compounded for larger cetaceans such as orcas, where fatalities among trainers have also fueled public backlash. Overall, the harm done to the captive animals outweighs any indirect potentially positive considerations such as promoting a broader public understanding of and appreciation for these top predators. In this vein, the long-term benefits of the expensive and much touted dolphin therapy programs for children with disabilities have largely been debunked: most studies do not support significant longer-term developmental progress from the dolphin interactions, whereby considerably simpler therapy sessions with dogs, horses or other animals have the same short-term benefits.

Clegg, I.L.K., Delfour, F. 2018. Can we assess marine mammal welfare in captivity and in the wild? Considering the example of bottlenose dolphins. *Aquatic Mammals* 44(2): 181–200. https://doi.org/10.1578/AM.44.2.2018.181

Salgueiro, E. et al. 2012. Effects of a dolphin interaction program on children with autism spectrum disorders – An exploratory research. *BMC Research Notes* 5: 199. https://doi.org/10.1186/1756-0500-5-199

67. **Fish is supposed to be such a healthy diet, and you want me to stop eating all my favorite fish?!**
    - If we eat the last ones, no one will ever enjoy them again.
    - If it's a fisher's right to catch fish, then why shouldn't it be my right to demand that the fish stay unscathed in their habitat?
    - Yes, fish is a healthy source of protein and you should enjoy it regularly (if you're not a vegetarian . . .). The fact, however, is that

wild-caught fish are "wildlife" and that some species and populations are endangered. The catchword is overfishing. Several types/definitions of overfishing are recognized, one of which is putting ever more effort into catching a target species, but not catching correspondingly more. Basically, it means catching fish faster than that population can replenish itself. This reflects a mixture of ever more sophisticated fishing techniques, setting official quotas too high (to satisfy the fishing industry), ignoring or failing to adequately control the quotas in place, paying subsidies to fishers, and illegal fishing. It also reflects an ever-increasing global human population and demand. The result is that, while the global fisheries catch rose consistently in the early 20th century (and once promised to solve world hunger), catches have now flattened out. One-third of the world's marine fish stocks are overfished, 60% maximally fished. Aquaculture/mariculture is only a partial solution (see entry Nr. 89). This has already led to complete fishery closures (for example cod in eastern Canada) and to various "fishing wars." The advice to the environmentally minded consumer is therefore to know which species/populations are endangered and to avoid these whenever possible. The relevant lists are typically produced by NGOs and the categories may be color-coded such as red (avoid), yellow (moderate consumption), and green (recommended) species. The classic example for "red" is the bluefin tuna, whereby scarcity and hype have reached such a level that a record $3.2 million was paid for a single fish at a Japanese auction in early 2026. Enjoy fish – maybe some of them should be freshwater species and/or from certified fish farms.

Clovis, H.I.A., Simon, A.M. 2024. Understanding overfishing: A literature review. *Asian Journal of Fisheries and Aquatic Research* 26(1): 61–71. https://doi.org/10.9734/ajfar/2024/v26i1727

https://ourworldindata.org/fish-and-overfishing

https://www.smithsonianmag.com/smart-news/massive-bluefin-tuna-the-size-of-a-motorcycle-sells-for-13-million-at-a-japanese-fish-market-180985799/

68. **Hey, it's only a flimsy plastic bag. It can't possibly be harmful (Figure 7).**

Figure 7 Whether sea turtles and other marine organisms swallow plastic bags because they confuse them with jellyfish (unlikely) or because everything floating in the sea was always worth nibbling at, the result is the same: slow death.

- It might be flimsy, but if you put it tightly over your head, can you still breathe?
- Fish, birds, dolphins, and most other marine life can't deal with plastic bags any better than children under the age of 3.
- Worldwide, nearly *two million* single-use plastic bags are used every *minute*. Only a few are recycled, and this type of bag makes up 5% of all items collected in annual coastal clean-ups. Sea turtles chomp on them, and their throats are designed to ensure that slippery prey such as jellyfish can't escape, so it's a one-way trip to blocking the digestive tract and a slow death by starvation. Plastic bags have also been found in a large range of marine mammals, and plastic, including remnants of bags, has been found in virtually every seabird species ever examined. If you don't find that disconcerting, plastic bags pose a threat to humans as well. At sea, many an outboard motor has been destroyed by a snagged bag that blocks the cooling water intake. This can cause the engine to self-destruct within seconds, leaving you adrift at no doubt the most inopportune time (storm coming up, darkness approaching, no more beer in the cooler).

Lenzi, J. et al. 2016. Plastic ingestion by a generalist seabird on the coast of Uruguay. *Marine Pollution Bulletin* 107(1): 71–76. https://doi.org/10.1016/j.marpolbul.2016.04.016

Rodríguez, Y. et al. 2023. Cetaceans playing with single-use plastics (SUPs): A widespread interaction with likely severe impacts. *Marine Pollution Bulletin* 194(Part A): 115428. https://doi.org/10.1016/j.marpolbul.2023.115428

Stachowitsch, M. 2019. *The Beachcomber's Guide to Marine Debris.* Cham, Switzerland: Springer. 358 pp.

69. **I paid a lot of money for this night-time sea turtle nesting beach tour, and the customer is king: don't tell me that my flashlight is a problem and that I can't use a flash when taking photos! (Figure 8).**

Figure 8 Light pollution is a recognized threat to wildlife (and to human well-being). One lightbulb can misorient turtle hatchlings: they run inland toward the light rather than into the water. Covering this nesting beach with dense furniture and rugs raises my blood pressure and anarchistic bent!

- Would you like to have people using flashes while you are executing a key life function?
- Sometimes it's good to be kept in the dark.
- Sea turtles dig the dark.
- Sharing (those photos on social media) is not necessarily caring.
- Normally, taking photos is the one activity that tourists encountering wildlife are welcome to do (but see entry Nr. 73 regarding underwater photographers). Caveats are in place, however, for photography on beaches during the sea turtle nesting season. Light is a key factor determining sea turtle behavior, especially that of hatchlings on their way to

the sea. They can become misoriented by artificial light sources on or behind the beach (because they normally move away from the darker hinterland toward the "glistening" water). Fortunately, light is among the easiest forms of pollution to stop: simply turn off the switch. This is also one of the reasons why using camera (and cell phone) flashes or spotlights, screens, headlamps, or flashlights is forbidden during the nesting and hatching season on most beaches worldwide. If you do take photos, make sure your flash setting is disabled. Someone in the group with a more sophisticated camera will no doubt be willing to forward the best shots to you. And remember, all sea turtles are protected (almost) everywhere, and that encompasses harassment.

https://awesomeocean.com/conservation/cell-phones-harm-sea-turtles/

https://seaturtlespacecoast.org/lights-out-please-its-nesting-season/

70. **Swimming with dolphins/whales is such a more intense experience than watching them from the deck, so let me jump in with my mask and snorkel.**
    - How would you like people lining up inside your home to watch your daily routine close-up?
    - Whale watching has become a huge, billion-dollar industry. It is much touted as an alternate, non-lethal "use" of whales (i.e., as opposed to whaling). At the same time, this activity is a recognized threat to whales and dolphins. The problems begin with collisions with whale watching vessels, mostly on the higher-speed journey to and from the whale watching site. To reduce disturbance, minimum distances are prescribed, typically 100 m for a single boat, more if a calf or when two or more boats are present. Approach angles and the duration of encounters are also regulated. The International Whaling Commission's Whale Watching Handbook is the go-to tool for managers, regulators, and tour operators. How to recognize disturbance? When the animals change course or speed, when they dive and do not resurface at their normal intervals, when they cease feeding or resting, when the group splits up, when they stop communicating, etc. Adhering to established guidelines would normally mean that you should never be close enough for closer contact with the animals. If the tour operators bring you close enough to jump in next to them, they have probably broken the rules by pursuing, heading off, or encircling the cetaceans. Consider reporting them to the authorities upon returning to port. So-called "swim-with" programs are generally opposed by most relevant organizations. This is especially the case when top predators or large whales are involved, which can pose a threat to humans in the water (as would closely approaching any wildlife weighing up to many tons).

Hale, S.E. 2025. Snorkelling with orcas (killer whales) in Skjervøy, Northern Norway: Ambivalent encounters in a crowded tourism space. *Environment and Planning E* 0(0). https://doi.org/10.1177/25148486251319672

https://wwhandbook.iwc.int/en/

71. **These coral reef fish are clearly hungry, so I'm taking along a bag full of bread.**
    - Eating the wrong food will kill a fish much faster than a human.
    - Most fish spend a large part of their day foraging, seeking any opportunity for a meal. Most also have very specialized diets, which is often reflected in the particular shape and structure of their mouthparts. No one can accurately simulate that diet with leftovers brought in a plastic bag. This misguided practice is generally of two kinds: tourism staff seeking to attract fish for their clients (e.g., shark feeding attraction), and divers seeking to get up close, for example, for photographs. The impacts on the fish include changes in species distribution and behavior, negative health effects, and increased predation of some species. Moreover, the fish may become habituated and approach people in the water, including the next spearfisher. Again, if all these arguments fail to sway you, then such activity also poses a risk of injury to tourists: some fish with sharper teeth may confuse your fingers with that hotdog fragment. Importantly, feeding smaller fish is the best way to attract much larger and hungrier fish with many even longer and sharper teeth.

Patroni, J. et al. 2018. Feeding wild fish for tourism – A systematic quantitative literature review of impacts and management. *International Journal of Tourism Research* 20(3): 286–298. https://doi.org/10.1002/jtr.2180

72. **These corals are rock hard – I can't see why touching or sitting on them is such a problem.**
    - Your head is hard – do you want passers-by to poke their fingers into it, grab and sit on it, cut off a souvenir tuft of hair, or bend your ear back and forth?
    - Yes, the skeleton of many coral species may seem to be hard, but the living polyps themselves are soft and form only a very thin surface layer on the colony. Squashing or holding onto a colony can injure and kill the polyps. Moreover, many branching colonies are actually quite fragile, and grabbing or brushing against them can break off pieces. Any such injury exposes the colonies to pathogens. Even if colony repair is successful, it can be quite slow. Beyond the damage done by recreational divers (see entries Nr. 73, 86), spear fishers touched corals at a rate of more than once per minute and caused physical damage at a rate

0.5 times per minute. Massive corals were most frequently touched, and branching corals were most frequently damaged. Both problems occurred mainly through fin kicks, spearfisher body contact, and the speargun. If you happen to be more worried about yourself than the corals, the exposed parts of the coral skeletons tend to be very sharp, and many a skin layer and neoprene diving suit of inattentive snorkelers and divers have been slashed.

Giglio, V.J. et al. 2017. Behaviour of recreational spearfishers and its impacts on corals. *Aquatic Conservation: Marine and Freshwater Ecosystems* 28(1): 167–174. https://doi.org/10.1002/aqc.2797

73. **To get a good photo of that fish, I simply have to attract it with food or push aside those sponges/soft-coral branches, etc.**
    - The next time someone swims by with a speargun instead of a camera (and food), how do you think those diver-fed fish will react?
    - Fish may have relatively simple brains, but – as every aquarium-holder knows – they can learn important things fast (like who brings the food and when). This leads to habituation, which is rarely a positive habit for wildlife exposed to the public. Habituation also clearly poses a threat to humans when larger predatory species become accustomed to people and don't get the goodies they expect (or mistake you for the goodie). Importantly, underwater photographers tend to cause the most damage to reefs. Beyond feeding fish (to avoid the typical "tailfin-only" perspective of fleeing individuals), they come into contact and break more corals than "normal" divers, that is, either male or female divers without cameras. This is because they are either distracted from carrying out their normal diving skills, wish to stabilize themselves in the current by holding onto or kneeling in the corals, or sometimes move or remove organisms and structures that are in the way of a perfect picture. Some guidelines: don't touch any animals or move them to a "better" background; don't move or break corals to get a clear shot; don't push the camera too close to get a close-up shot; don't repeatedly take flash photos of the same animal; don't use excessively bright focusing lights; don't neglect your diving (buoyancy) skills; and stay aware of your surroundings. You don't want to be fully concentrated on taking snapshots of that cute anemone fish while a shark is circling ever closer from behind, or an irate moray eel is about to clamp its jaws on your ankle or some other more sensitive body part. Of course, photographer-related damage goes beyond coral reefs to include other ecosystems such as sediment bottoms and their many inhabitants.

De Brauwer, M. et al. 2018. Time to stop mucking around? Impacts of underwater photography on cryptobenthic fauna found in soft sediment habitats.

Journal of Environmental Management 218: 14–22. https://doi.org/10.1016/j.jenvman.2018.04.047

Rouphael, A.B., Ingli, G.J. 2001. Take only photographs and leave only footprints? An experimental study of the impacts of underwater photographers on coral reef dive sites. *Biological Conservation* 100(3): 281–287. https://doi.org/10.1016/S0006-3207(01)00032-5

74. **Sea urchins are a sign of pollution, so my friends and I just killed a few hundred of them underwater with a hammer.**
    - So, what is your expert opinion about how many sea urchins it's OK to kill?
    - So, what other icky or annoying animals should be wantonly smashed?
    - Killing animals is, in fact, one type of acknowledged wildlife management, but typically only as a last resort and after broad approval by higher-ups – and then conducted by professionals. Moreover, such official killing typically focuses on introduced alien species that pose a severe threat to the native fauna or the overall ecosystem. Sea urchins have actually been known to exhibit population explosions that harm the surrounding growth such as kelp beds due to overgrazing. Such explosive growth probably reflects poorly understood and compromised food webs and may involve pollution or the prior removal or decimation of sea urchin predators. It is highly unlikely that large-scale sea urchin invasions can be combatted by inexperienced visitors killing individuals underwater. Importantly, mass mortalities of sea urchins also occur, usually after natural heat stress or freshwater incursions. In some places, humans also harvest large numbers of sea urchins for food, which can impact the population structure of the respective species. No one should kill any organisms at scale in the marine environment unless trained to do so within the frame of an officially authorized eradication program. This, for example, may be the case in mass outbreaks of crown-of-thorns seastars, which endanger the livelihoods of local dive centers and pose an immediate threat to the survival of entire coral reefs.

Hendler, G. 2013. Recent mass mortality of *Strongylocentrotus purpuratus* (Echinodermata: Echinoidea) at Malibu and a review of purple sea urchin kills elsewhere in California. *Bulletin of the Southern California Academy of Sciences* 112(1): 19–37. https://doi.org/10.3160/0038-3872-112.1.19

Pratchett, M.S. et al. 2019. Culling crown-of-thorns starfish (*Acanthaster cf. solaris*) on Australia's Great Barrier Reef: Rationale and effectiveness. *Australian Zoologist* 40(1): 13–24. https://doi.org/10.7882/AZ.2018.021

75. **These whales/dolphins beached themselves. If they are so stupid, they deserve to die.**

- So, all car drivers who took a wrong turn and ended up in a dead end or in a very bad neighborhood should die?
- Every year, whales and dolphins end up stranded on beaches around the world. The potential causes of this phenomenon have been widely discussed for decades, but stupidity is certainly not one of the explanations that has ever been advanced. Stronger arguments have been made for impairment due to parasites, disease, injuries, loud underwater noise, fisheries interactions, and the mismatch between geomagnetic lines in the water (along which some cetaceans apparently orient themselves) and coastline morphology. This has led the International Whaling Commission to launch a Strandings Initiative led by a Strandings Coordinator and supported by an Expert Panel and a Steering Group. The program is designed to provide capacity-building training on stranding response around the world. Beyond collecting data (including performing necropsies), the animals can sometimes be saved. If not, then animal welfare considerations can mean that trained personnel are required to kill the animals with special apparatus or injections.

Coombs, E. et al. 2019. What can cetacean stranding records tell us? A study of UK and Irish cetacean diversity over the past 100 years. *Marine Mammal Science* 35(4): 1527–1555. https://doi.org/10.1111/mms.12610

https://iwc.int/management-and-conservation/strandings/strandings-initiative

76. **These animals are occupying half the beach and shitting all over the place. Why should we have to put up with that?**
    - Would you like other people to determine where and when you were allowed to sit or shit?
    - You should really know a lot about natural biological processes before you change something that has been successfully going on for millions of years.
    - Many marine mammals spend considerable time on beaches, for example to rest, warm themselves, mate, or give birth to and nurse their young. Many populations were heavily decimated in the past, and strict protection has led to population increases in certain areas – accompanied by correspondingly larger numbers seeking suitable shorelines. In some case, this has meant more large marine mammals on beaches. It can also mean animals occupying the docks, jumping on moored private boats, and using other artificial structures and barriers that we have erected along the coast. This can be disconcerting to visitors on beaches because the animals potentially pose a threat, especially because people fail to recognize the need to keep a safe distance and often succumb to cell phone camera mania. Large animals on docks can hamper human

activity and compromise private boats, which are no doubt the third-most valuable possessions of most boat owners (after their homes and cell phones). However, the laws in many countries (in the US the Marine Mammal Protection Act or the Endangered Species Act, for example) include very strict regulations against harassing the animals; those regulations include various attempt to scare them off or otherwise shoo them away. The number of negative interactions with humans (and their dogs) can be high and require animal rescues. In some cases, "nuisance" animals are killed, like the walrus Freya in Norway that climbed onto (and sank) docked boats. The reason given by the Directorate of Fisheries there, however, was public safety based on irresponsible onlooker behavior. Some guidelines: keep your distance, keep dogs on a leash, be a responsible photographer, don't intervene to "help" the animals, and call the appropriate authorities if problems arise.

Levin, A., Vincent, S. 2023. The life and death of Freya the Walrus: Human and wild animal interactions in the Anthropocene era. *Animals* 13: 2788. https://doi.org/10.3390/ani13172788

https://www.marinemammalcenter.org/marine-mammal-protection-act

77. **It's a wonder of nature to see how pufferfish keep inflating themselves when we poke them (Figure 9).**

Figure 9 Harassing a puffer- or porcupine fish is just one of the many unacceptable behaviors that some divers display. Is sessile or slow-moving wildlife fair game for touching to see how it reacts? Save poking for your cell phone display! (Image from Shutterstock).

- Every time someone pokes you in the eye, it waters. So much fun!
- Almost everyone has misbehaved at some point in time on land, but why misbehave underwater as well?

- This is a classic example of divers harassing marine wildlife. To protect themselves, pufferfish and the related porcupine fish can inflate their bodies several-fold to avoid being swallowed. This is one of the most iconic non-evasive predator defense strategies in nature and involves rapidly gulping water into their distensible stomach. Triggering this mechanism "for the fun of it" or to take a photograph (because the fish then becomes less mobile) seriously stresses the fish and requires it to waste crucial energy resources. And don't forget, they can bite you, and some are among the most toxic vertebrates on the planet, causing human fatalities if improperly gutted and cooked ("fugu"). Any form of harassment that causes an animal to change its behavior or take evasive action represents a waste of that animal's energy. It may then be less fit or able to engage in other crucial, life-sustaining activities.

McGee, G.E., Clark, T.D. 2014. All puffed out: Do pufferfish hold their breath while inflated? *Biology Letters* 10(12): 20140823. https://doi.org/10.1098/rsbl.2014.0823

78. **It was amazing to see how our bubbles collected on the roof of the cave we were diving in (Figure 10).**

Figure 10 Leave only your bubbles is an often-heard guideline for scuba divers. Unfortunately, rising air bubbles from divers in this cave have blocked the mouth opening of each of these coral polyps. Some are already dead and covered by a gray sponge; the others will probably suffer the same fate.

- And the animals growing on that roof are now hanging in the air?
- "Hot air" is annoying enough on land; it can be lethal underwater.

- One of the standard mantras in scuba diving circles is "leave nothing but bubbles," which is perhaps the marine equivalent of the terrestrial "leave nothing but footprints" (but see entry Nr. 40). Unfortunately, even bubbles are now a recognized visual and audible physical disturbance in aquatic ecosystems. Accordingly, they typically repel fish, which is why some fish photographers and researchers conducting fish counts or documenting fish behavior use closed-circuit rebreather systems that emit no bubbles. Bubbles also accumulate under overhangs or on cave roofs, whereby the organisms growing there end up in an air space and die. They have even been implicated in partial cave collapses. Rising bubbles have also been reported to disturb plankton communities. While these effects are probably negligible in most cases, they do underline that we leave traces wherever we go. The take-home message? Perhaps we should consider whether we really need to go everywhere (dive inside every cave) and, if so, how we can do it with as little impact as possible.

Girihagama, L. et al. 2015. Bubble-induced cave collapse. *PLoS ONE* 10(4): e0122349. https://doi.org/10.1371/journal.pone.0122349

Lopes, K.H. Jr. et al. 2019. Effects of SCUBA bubbles on counts of roving piscivores in a large remote marine protected area. *PLoS ONE* 14 (12): e0226370. https://doi.org/10.1371/journal.pone.0226370

79. **What do you mean I can't drive across these beach dunes in my 4-wheel drive? (Figure 11).**

Figure 11 The tire tracks of vehicles can scar sandy areas by breaking the sensitive surface crust, promoting erosion. Off-road racing vehicles in the US desert, for example, left tracks that are still visible decades later. On beaches, vehicle ruts pose a threat to sea turtle hatchlings by compacting the sand above the nests, which hinders emergence, but also by "capturing" and misorienting those on their dash to the sea.

- If you promise not to leave any ruts or run over any plant, go ahead.
- Why can't you just go cruising for burgers in daddy's new car? (Frank Zappa)
- The rear parts of beaches and beach dunes are dynamic transitional zones between the beach proper and the hinterland. They are habitats in their own right, typically with a unique vegetation and a fauna that differs from that of the beach itself. They are recognized as being very sensitive to disturbance, and many have been destroyed in the past to make way for tourist and other facilities. Accordingly, remaining dune landscapes are now often under strict protection. Even visitors crossing them on foot in larger numbers every day can damage them. One solution is to construct elevated walkways to enable an environmentally friendly access to the beach. Beyond conservation considerations, coastal dunes have a strong aesthetic appeal and provide an important service by buffering coastal communities against storm waters. If walking across them is already an issue, then driving off-road vehicles (ORVs) over them as recreation or sport is the environmentally most damaging form of direct dune use and a real no-no. ORVs on dunes crush vegetation, reduce animal species abundance and diversity, disturb and kill birds (including their nests), accelerate erosion, and are sometimes associated with camping along with its inevitable waste production. The recreational ORV lobby is powerful, and people pay lots of money for those "wheels," in part driven by pervasive media and manufacturer advertising on the purported outdoor experience they enable. Don't believe the hype! Sand dunes shouldn't be part of that faulty narrative.

  Schlacher, T.A. et al. 2025. Adverse impacts of off-road vehicles on coastal dune vegetation are widespread, substantial, and long-lasting: Evidence from a global meta-analysis of sandy beach-dune systems. *Estuarine, Coastal and Shelf Science* 312: 109038. https://doi.org/10.1016/j.ecss.2024.109038

80. **How can you expect people to visit my beach bar at night if I can't put up neon signs and play loud music?**
    - How would you like neon lights and loud music in your bedroom every night?
    - Sure, and while you're at it, don't forget nightly fireworks!
    - This is a classic example of the conundrum when people and wildlife vie for the same space. Perhaps that phraseology is a key to

the issue. For people, in this case, for a commercial enterprise, it is just that – a space like any other. For the fauna there, however, it's the one and only habitat. Who do you think should have priority? What's better at night for the wildlife there: a dark or an illuminated beach? The effect on sea turtles is perhaps best studied and is a typical disturbance that nesting beach monitoring teams encounter in many parts of the world. The chance for hatchlings to successfully reach the sea upon emerging from their nests at night under bright lights is strongly compromised: in fact, one lightbulb or even a glowing beverage vending machine is sufficient to misorient them. In many cases, the business itself may be skirting legality in the first place: it may be erected on a protected beach or on a section of beach closed to commercial enterprises (e.g., too close to the water), or have exceeded the number of tables and chairs permitted, etc. Often, the prospects of profit override any fear of penalties. And, as almost everywhere, jobs tend to outweigh any conservation arguments. Solving this dilemma requires sensitivity and tact – but also a good measure of knowledge and perseverance on the part of biologists, both toward the owner/operator and the relevant authorities. And if the effects of night-time noise and lights on nature fail to convince you, you can be selfish and consider the effect on your sleep in your beachside hotel room.

Bird, B.L. et al. 2004. Effects of Coastal Lighting on Foraging Behavior of Beach Mice Efectos del Alumbrado Costero sobre el Comportamiento de Forrajeo de Ratones de Playa. *Conservation Biology* 18(5): 1435–1439. https://doi.org/10.1111/j.1523-1739.2004.00349.x

Lorne, J.K., Salmon, M. 2007. Effects of exposure to artificial lighting on orientation of hatchling sea turtles on the beach and in the ocean. *Endangered Species Research* 3: 23–30. https://doi.org/10.3354/esr003023

Sari, D. et al. 2014. Assessment of noise pollution sourced from entertainment places in Antalya, Turkey. *In: Inter-Noise and Noise-Con Congress and Conference Proceedings* (Vol. 249, No. 7, pp. 1072–1077). Institute of Noise Control Engineering.

81. We're helping the sea turtles by removing the beach umbrellas at night and sticking them back in the sand in the morning (Figure 12).

Figure 12 Every square inch/centimeter of the sea turtle nesting area on this beach has been monopolized for commercial profit. This is total disregard for species protection in an official Mediterranean "Special Protected Area."

- That strategy is under the umbrella of misinformation.
- Close but no bananas!
- That isn't the optimal nesting beach management option. It's true that all beach furniture (sunbeds, chairs, tables, umbrellas) should be removed from nesting beaches in the evening. This is because such objects can obstruct the crawls of adult female turtles when they emerge from the sea and search for a suitable nesting spot. Such furniture can be returned in the morning. The problem, however, is recognizing where a turtle may have dug a nest overnight: you don't want to put furniture on top of a nest (the eggs need the sun's heat to develop). In the case of umbrellas, you certainly can't risk sticking them into a nest: one or more perforated eggs can ultimately rot the entire clutch. Accordingly, one recommendation is to use permanently positioned umbrella stands or use umbrellas that fasten onto other furniture. Another is to ban the use of private sun umbrellas. A more overarching approach is to ban all beach furniture from the nesting zone itself. This still leaves plenty of room for beach-goers: they can settle down in front of or behind that

zone (which should be clearly designated). And, of course, it's helpful if nests are marked by nesting beach monitoring teams.

https://www.widecast.org/conservation/threats-and-solutions/obstacles-on-the-beach/

https://myfwc.com/research/wildlife/sea-turtles/threats/beach-furniture/

82. **Why don't you catch the sea turtle hatchlings at night and release them after breakfast so we can all see them?**
 - Being born was never meant to be a show for tourists.
 - What do you think those baby turtles are going through piled up in a bucket overnight?
 - Sea turtle hatchlings, which in most species emerge at night, are sometimes collected if the unfavorable position of the nest, barriers on the beach, or artificial lighting make it unlikely that they will be able to reach the sea on their own. These hatchlings are typically released on the same night at a nearby, more promising location. In situations where the eggs from natural nests have little chance of producing hatchlings (for example, the nest is in the wave swash zone) or the hatchlings are later unlikely to reach the surface or the sea (nest on a compacted dirt road or very far from the waterline), sea turtle workers may immediately relocate the freshly deposited eggs to a fenced-off, so-called hatchery. If, many weeks later, multiple nests hatch on the same night at such a hatchery, then the nesting beach monitoring team may not have had enough time to release them all on the beach that night. In such cases, tourists who get up early enough may have the luck to experience this crucial phase in a sea turtle's life during the team's morning shift. Sometimes such releases are coordinated with nearby hotels that might sponsor the hatchery or the program. Retaining emerging hatchlings solely for tourists on the following day, however, is generally frowned upon. It's an unfavorable disruption of a critical phase in the turtles' life cycle. Moreover, releases involving the public generally require extra effort and personnel to keep enraptured visitors in check. Many want to hold hatchlings; rarely does someone not wish to take close-up photos. The tiny turtles can easily be stepped on in the commotion. Sometimes local residents see the hatchlings as "their" animals and insist (especially to foreign teams) on hand-releasing them themselves. One way we successfully used to ensure that visitors stepped back or instantly "dropped" hand-held hatchlings was to utter the magic words "microbes" or "pathogens" with an appropriately horrified facial expression. Generally, sea turtle workers will try to accommodate visitors because of the positive environmental education

effect, but this is not always possible. Of course, donations will almost always be welcomed (see also entry Nr. 207).

Budiantoro, A. et al. 2019. The sustainable ecotourism potential development with special reference to olive ridley sea turtle (*Lepidochelys olivacea*) along Bantul beaches, Indonesia. *IOP Conference Series: Earth and Environmental Science* 391: 012069. https://doi.org/10.1088/1755-1315/391/1/012069

Tisdell, C., Wilson, C. 2005. Do open-cycle hatcheries relying on tourism conserve sea turtles? Sri Lankan developments and economic–ecological considerations. *Environmental Management* 3: 441–452. https://doi.org/10.1007/s00267-004-0049-2

83. **I saw these poor little sea turtles crawling around on the beach in the dark so I collected them and threw them into the sea the next morning.**
   - What other wild animals would you collect and keep overnight in your hotel room?
   - How long do you think it's OK to delay one of the most critical phases in a sea turtle's life?
   - One of the main mantras regarding tourists and wildlife is "don't intervene." If you believe there is a problem, then contact the appropriate authorities. In the case of sea turtle nesting beaches, a monitoring team is often present on site and willing and authorized to take the necessary action. Another version of that mantra is "things should progress as naturally as possible." Most untrained people are poorly equipped to determine whether a situation is proceeding normally or not. In the case of sea turtles, the dash of hatchlings to the sea after emerging from the nest is a short and crucial phase in their lives. They use their energy reserves to reach the sea and then quickly swim to safer waters offshore. If hatchlings are collected (for example because they are crawling in a direction away from the water or they are somehow trapped), the best practice is to release them that same night. This can help prevent hatchling exhaustion, dehydration, overheating, depletion of the energy reserves, injury, and mortality. With increasing time of captivity (and keeping in seawater tanks), their activity and orientation upon release may decrease. Of course, handling no doubt causes stress and should be avoided with any wildlife. Importantly, the hatchlings of most species emerge at night when their orientation is based on subtle light differences, the sand surface temperatures are not yet lethal, and most predators on land and in the water are asleep. Releasing them in the morning is therefore potentially problematic, and releasers may well attract and have to deal with unruly onlookers.

Albrecht, M. 1997. Activity and orientation of sea turtles in captivity during the first 100 hours after hatching. *In: European Research on Cetaceans:*

Proceedings of the Annual Conference of the European Cetacean Society (Vol. 11, p. 303). European Cetacean Society.

Phillot, A.D. 2023. Holding and headstarting sea turtle hatchlings. *Indian Ocean Turtle Newsletter* 38. https://www.iotn.org/iotn38-03-holding-and-headstarting-sea-turtle-hatchlings

84. **I'm a skilled boater and can drive fast though this sea cow habitat; after all, they can hear, and they're used to waves in the ocean, aren't they?**
    - So you have x-ray eyes and can see where they all are just below the murky surface?
    - Is skilled the same thing as experienced and mindful?
    - Boating (and larger ocean vessels) pose a problem to all air-breathing wildlife that spend time on the surface, including whales and dolphins, sea cows (manatees, dugongs), and sea turtles. In Florida, for example, the manatee mortality and injury rates, chiefly from boat propellers, once rose by 25% per year. This prompted a public education and awareness campaign and, ultimately, various speed restrictions (e.g., "idle and slow" (ca. 5–7 mph = 8–11 kmh) in restricted areas). The expert conclusion on avoiding collisions and reducing the severity of injuries in all cases is slower speeds. For example, vessels 65 feet and greater in length near key port entrances and in right whale hotspots along the U.S. eastern seaboard are required to reduce speeds to ten knots or less. In shallower or semi-enclosed waters – whether marine, brackish, or freshwater – one solution has been to declare "no wake zones." This strategy enables recreational captains to better spot and avoid animals, and the slower speeds also help to prevent resuspending the sediments that cloud the waters. It is also an important step in reducing the erosion of sensitive shorelines along with their vegetation due to artificially created waves. In the case of sea cows, which feed on seagrasses, inattentive boating in shallow waters can scar the seagrass beds, promoting erosion, fragmentation, and even loss of the meadows, severely damaging the entire ecosystem. Although most aquatic animals can hear quite well in water (although some, like manatees, hear certain frequencies well, others poorly), few are adapted to react to high-speed boats, not to mention crazed amateurs testing the limits of their freshly rented jet-skis.

Laist, D.W., Shaw, C. 2002. Preliminary evidence that boat speed restrictions reduce deaths of Florida manatees. *Marine Mammal Science* 22(2): 472–479. https://doi.org/10.1111/j.1748-7692.2006.00027.x

Shackley, M. 1992. Manatees and tourism in southern Florida: Opportunity or threat? *Journal of Environmental Management* 34(4): 257–265. https://doi.org/10.1016/S0301-4797(11)80002-2

Silber, G.K. et al. 2012. *An assessment of the final rule to implement vessel speed restrictions to reduce the threat of vessel collisions with North Atlantic right whales*. NOAA technical memorandum NMFS-OPR-48. https://repository.library.noaa.gov/view/noaa/4207_DS1.pdf

85. I'm just a snorkeler, you should be pestering those crazy divers, not me (Figure 13).

Figure 13 Inexperienced snorkelers can do as much damage to a reef as rambunctious scuba divers. Fragile corals break, and surviving colonies must expend considerable energy (by producing mucus) to remove the resuspended sand landing on them.

- Let me see your snorkeling certification.
- Then why are you standing on the reef?
- Contrary to what many might believe, snorkeling need not necessarily be more environmentally friendly than diving. This is primarily because many vacationers may be borrowing or renting snorkeling equipment for the very first time and need all their wits just to keep themselves from drowning. This means having little remaining capacities free to stay clear of and go easy on corals or other marine fauna and flora. Snorkelers are often unaware of or entirely underestimate their contact rate with underwater life, which one study determined was more than one damaging behavior every five minutes. Most were fin kicks:

inexperienced snorkelers typically fail to factor in their overall length with flippers on and either sheer off coral colonies or stir up the sand and cloud the water. Corals must expend considerable energy to remove such sediments raining down on them (by producing abundant mucus and sloughing the particles off). This type of damage was followed by sitting, standing, or kneeling on the reef. Another study reported 33% more contacts and 70% more damage to benthic reef organisms than by scuba divers, whereby the contact rate increased more than five-fold when they interacted with sea turtles. Pre-trip information (and signing a written pledge) considerably reduced such behavior with a two-fold benefit: it helps ecotourism facilities to sustain the reef and to safeguard their economic livelihoods. A short cartoon video (silent and therefore "international") for visitors to a Maldives resort has also been tested.

https://www.youtube.com/watch?v=1Y8onPPPfCM

Saliba, B.M. et al. 2023. Interactions of divers with reef biota are more frequent among snorkelers than scuba divers and increase during sea turtle watching. *Tourism in Marine Environments* 17(4): 249–263. https://doi.org/10.3727/154427322X16710976626847

Webler, T., Jakubowski, K. 2016. Mitigating damaging behaviors of snorkelers to coral reefs in Puerto Rico through a pre-trip media-based intervention. *Biological Conservation* 197: 223–228. https://doi.org/10.1016/j.biocon.2016.03.012

86. **I'm a certified diver; you should be pestering those hordes of crazy snorkelers, not me (Figure 14).**

Figure 14 Every diver must get into and out of the water. This often involves large groups wading across the shallow reef top – a major cause of damage and a special form of "trampling."

- You probably do the same amount of damage as snorkelers, only a bit deeper.
- "Scuba divers do it deeper." (popular T-shirt slogan).
- Yes, scuba diving is a wonderful thing, and experiencing the beauty of the underwater world can kickstart your interconnectedness with nature and conservation awareness, much like the awe and humbleness that overwhelms astronauts when viewing the planet from above. At the same time, scuba divers are a main cause of local physical (anthropogenic) damage in coral reefs. The degree of damage is correlated with the number of divers/dives, with some hotspots registering hundreds of thousands of dives per year. Coral health declines (and disease prevalence increases) in high-use versus low-use sites. The problems start when entering the reef, which often involves walking across the reef top in full gear. Nearly ¾ of all divers make contact with the reef when diving, averaging a dozen contacts per dive (sequence: flippers, hands, knees, diving gauges). Men cause more coral breakage than women, with underwater photographers having the worst record (see entry Nr. 73). Sediment resuspension (and deposition on corals) is also an important diving-related impact. Strategies to counter such damage include:
  - pre-dive briefings;
  - a check dive to determine the divers' buoyancy skills;
  - fixed moorings rather than anchoring (if access is by boat);
  - elevated walkways or jetties over the reef top (if entry is on foot);
  - a dive guide accompanying every group;
  - banning the use of gloves (see entry Nr. 87);
  - providing alternate underwater attractions adjoining the natural reef (boat wreck, practice platform, artificial reef structures);
  - special courses ("underwater naturalist");
  - limits to overall diver numbers; and
  - marked underwater (educational) trails to channel visitor flow.

Camp, E., Fraser, D. 2012. Influence of conservation education dive briefings as a management tool on the timing and nature of recreational SCUBA diving impacts on coral reefs. *Ocean & Coastal Management* 61: 30–37. https://doi.org/10.1016/j.ocecoaman.2012.02.002

Hasler, H., Ott, J.A. 2008. Diving down the reefs? Intensive diving tourism threatens the reefs of the northern Red Sea. *Marine Pollution Bulletin* 56: 1788–1794. https://doi.org/10.1016/j.marpolbul.2008.06.002

Sumanapala, D. et al. 2023. A review of ecological impacts from recreational SCUBA diving: Current evidence and future practice. *Tourism and Hospitality Research* 23(4): 564–577. https://doi.org/10.1177/14673584221112602

87. **I always take gloves with me so that I can hold on to the reef without a problem (Figure 15).**

Figure 15 The list of do's and don'ts in coral reefs is long. The rules are etched in stone as pictograms here to address visitors from around the world. "No gloves" is the first item in row 3.

- That might solve your problem, but not the reef's.
- Do you use gloves when touching other things you love?
- Scuba diving requires a lot of expensive technical equipment and lots of optional accessories, one of which is gloves. These are helpful if the water is cold (not an issue in coral reefs) or you are diving on a rusty wreck or along a rocky shore in the current and may need to protect yourself by briefly shoving off or holding on. In coral reefs, however, holding on can only mean grabbing corals, and that is guaranteed to damage them or, more typically, break the colonies. Moreover, wearing gloves considerably reduces the inhibition to pick up a souvenir, turn over a sea star, squeeze a sea cucumber, hold on to a sea turtle, poke a porcupine- or pufferfish, grab a nurse shark's tail, or otherwise harass marine life as many rambunctious divers are known to do. This is why gloves are typically forbidden by better diving centers in coral reef regions. This should be explained and controlled in pre-dive briefings. Good divers should be neutrally buoyant and never land on the bottom or touch anything.

Lin, B. 2021. Close encounters of the worst kind: Reforms needed to curb coral reef damage by recreational divers. *Coral Reefs* 40: 1429–1435. https://doi.org/10.1007/s00338-021-02153-3

88. **I only pick up/touch/poke/the sea life I know is harmless.**
   - You're touching a sore point.
   - So, you have the magic touch?
   - Well, then you probably know more than many marine biologists.
   - Why do we want to touch sea life? It's probably because our other four senses (smell, taste, hearing, sight) are dulled underwater. If you succumb to that urge, you should know that you are not protecting the species or their environment. Every animal you encounter in a reef or other marine habitat is in the exact position it needs to be in order to be either feeding, to protect itself or its young, to be camouflaged or resting. Touching, digging out or lifting up an invertebrate, whether it be a snail, crab, seastar, or sea cucumber, puts it in immediate peril. The same holds true for flushing out or doggedly pursuing a fish. Some predator is no doubt nearby and may even be following you closely to snatch that hapless prey as its next meal. At the very least, you are forcing the species to take defensive action or flee, and that means it is wasting valuable energy. Moreover, many so-called harmless species have protective mechanisms that can be unpleasant to humans (spines, stinging mechanisms, etc.). Some "harmless-looking" cone snails, for example, can inflict lethal injuries with their concealed, harpoon-like feeding apparatus: do you know which ones they are? If you are thinking about any such activity, you may well be wearing gloves to safeguard your pinkies – this whole issue wouldn't normally come up if you were diving with bare hands, where touching anything is risky. Better diving centers know why they forbid gloves. Be a responsible, environmentally aware visitor. Handling or even too closely approaching any wildlife is harassment, and any form of harassment is strictly prohibited and heavily fined almost everywhere. Save touching for your diving instruments, your diving buddy or approved "touch exhibits" in public aquariums.

Dogu, H. et al. 2011. Touch Exhibits for Aquatic Animals. Best Management Practices of Aquatic Animal Touch Exhibits. http://www.youtube.com/profile?user=DRJULIEFAGANSTUDENTS#p/u/0/GiowrPsMXd0

89. **Fish aquaculture/mariculture makes them bigger and better than in nature, and helps protect against overfishing the natural population.**
   - Define "better."
   - We haven't succeeded in making a better human being, so I don't think we can claim to have made a better fish.

- The world's fishing industry has reached and partially exceeded ("overfishing") its limits. This has led to a corresponding increase in mariculture operations to meet the ever-increasing demand for seafood. The only real advantages are commercial: mariculture creates jobs and produces constant yields, independent of any fishing season, quotas or extreme weather. This is balanced by a range of drawbacks:

  - fish are held in dense numbers, promoting stress, disease, and parasite infection;
  - exploding parasite populations can then impact the natural fish population;
  - this requires the use of antibiotics and other medications;
  - cultured fish therefore contain more chemical residues than the natural population;
  - cultured fish are sometimes genetically modified (faster growth, larger sizes);
  - cultured fish escape into the wild and can negatively affect the natural population (hybridization);
  - mariculture facilities may merely "raise" fish but require juveniles from the natural population (capture-based mariculture);
  - the food fed to the fish typically contains fish meal, which is processed from wild-caught fish;
  - mariculture facilities attract predators (e.g., sharks, porpoises) that can damage the pens/nets ("depredation") or become entangled themselves;
  - one ton of cultured fish defecates as much as one ton of cow; where does all that go?
  - feces and uneaten food rain down on the seafloor, typically promoting nutrient enrichment (eutrophication and resulting algal blooms) and causing "dead zones" that require moving the pens to a fresh site on a regular basis;
  - many mariculture facilities have been established on former mangrove forest areas that were felled specifically for that purpose; and
  - mariculture facilities often require large amounts of freshwater and thus compete with agriculture in arid zones.

  Most of the above drawbacks cannot be eliminated or circumvented. The take-home message? Mariculture will remain an important source of protein for human consumption but cannot replace fisheries, cannot be significantly expanded in the future, and cannot be marketed as being particularly environmentally friendly.

Lounas, R. et al. 2020. Towards sustainable mariculture: Some global trends. *Thalassas* 36: 447–456. https://doi.org/10.1007/s41208-020-00206-y

90. **Instead of catching a random fish by chance with a rod and reel, I can pick out the species and biggest fish while scuba diving with my speargun.**
    - Should the best fish be reserved for you or for the environment?
    - Harpooning can quickly decimate the larger fish in snorkeling depths. Extending that activity to much greater depths by using scuba can remove all the larger fish from the ecosystem. This is an environmental problem because fish size is often related to the ability of a species to maintain its population. On the one hand, a somewhat larger female can produce an order of magnitude more eggs than a slightly smaller individual. Targeting large females can therefore reduce the reproductive capacity of the population (see also entry Nr. 117). On the other hand, in many species the largest individuals are the only females or the only males present. In these cases, all smaller individuals either initially develop as females and then turn into males at a certain size (protogyny), or all smaller individuals are males and become functional females upon reaching a larger size (protandry). These are two forms of a phenomenon termed sequential hermaphroditism, whereby fish are the only vertebrates in which this occurs. For these and other reasons, spearfishing with scuba is prohibited in most regions of the world, whereby practiced free-divers today are capable of hunting in considerable depths. Other researchers, however, see a potential role of spearfishers in "inclusive participatory management," which could ostensibly represent a step toward "transformative sustainable development of marine recreational spearfishing." Hmmm. Perhaps "the thrill of the hunt" can also be achieved by replacing the speargun with an underwater camera, yielding a less lethal and more sustainable species protection and nature conservation outcome?

http://coraldigest.org/index.php/FishHermaphroditism

Lindfield, S.J. et al. 2014. Depth refuge and the impacts of SCUBA spearfishing on coral reef fishes. *PLoS ONE* 9(3): e92628. https://doi.org/10.1371/journal.pone.0092628

Sbragaglia, V. et al. 2023. A global review of marine recreational spearfishing. *Reviews in Fish Biology and Fisheries* 33: 1199–1222. https://doi.org/10.1007/s11160-023-09790-7

91. **You can't force me to use a mooring – I paid tons of money for this yacht so that I could be free to go and anchor wherever I want (Figure 16).**

Figure 16 Anchoring is just one of many potentially harmful environmental side effects of yachting (wake waves, bilge water pumping, improperly disposing of the "black water" in a boat's holding (poop) tank). Anchors and anchor chains crush and scar seafloor animal and plant communities. Multiply that damage times the estimated 30 million pleasure craft that anchor daily in shallow coastal waters worldwide.

- You mean your freedom is more important than intact shallow coastal ecosystems for everyone?
- You're leaving a terrible physical footprint on the seafloor several times larger than your boat every time your drop anchor.
- Your pleasure for one day or night, years of recovery for the ecosystem.
- Seafloors and their fauna and flora are under threat worldwide through human activities such as fishing and other commercial practices (e.g., bottom trawling, dredging, and probable future deep-sea mining). Anchoring by pleasure craft is also recognized as a major disturbance. This pertains especially to coral reefs and seagrass meadows, where heavy anchor chains can shear away, scour, and demolish entire habitats. Many shallow-water bays (e.g., in the Mediterranean) are completely occupied every day and night by multiple pleasure craft whose anchors and lines rip up the entire bottom, leaving lasting scars and never permitting recovery. Promising management practices include anchor-free zones, limiting the number of boats per anchoring site, and strict controls. Permanent mooring buoys anchored with a heavy weight to the bottom along a reef or installed in a bay by the authorities enable one or more boats to attach themselves without every captain having to drop anchor on a sensitive underwater habitat.

Jameson, S.C. et al. 2007. A quantitative ecological assessment of diving sites in the Egyptian Red Sea during a period of severe anchor damage: A baseline for restoration and sustainable tourism management. *Journal of Sustainable Tourism* 15(3): 309–323. https://doi.org/10.2167/jost719.0

La Manna, G. et al. 2015. The detrimental consequences for seagrass of ineffective marine park management related to boat anchoring. *Marine Pollution Bulletin* 90(1–2): 160–166. https://doi.org/10.1016/j.marpolbul.2014.11.001

92. I know this sculpture is glued together out of hundreds of colorful snails or other shells, but they were all picked up dead and dry on the beach, right? (Figure 17).

Figure 17 At first glance, maybe "clever" or "cute," but hundreds of snails were killed for this one item. Multiply that by hundreds of such items per curio shop and thousands of shops along the world's coastlines.

- No, the perfect, colorful shells glued together here aren't piled up by the millions on the beach.
- Such tacky sculptures are only one facet of the so-called curio trade. This includes many other "decorative" articles such as shark teeth and jaws, dried and inflated porcupine fish, dried seastars and seahorses, or dried and bleached corals. Please avoid purchasing such kitschy objects as mementos or souvenirs from seaside vacations, even if they are being passed off as "seashell art" or worked into jewelry. You are only perpetuating a cruel and unsustainable commercial exploitation of wildlife that currently encompasses hundreds of species of molluscs, many dozen species of coral, and an unknown number of sponges and other marine life. From another perspective, multiply the number of molluscs in that figurine by the number of figurines in the shop, then multiply that times the number of shops lining the coastline, in hotel complexes, and dotting the main shopping streets of coastal towns and cities. Astronomical! In many cases, the organisms involved are strictly regulated by international agreements such as the Convention on the International Trade of Endangered Species (CITES) and can be confiscated by border customs agents, who will no doubt also impose heavy fines. This once again highlights that species protection and nature conservation may be strictly regulated on paper, but that enforcement is often weak to non-existent. And it also highlights that you yourself can play a role in improving the situation.

http://www.oceansatlas.org/subtopic/en/c/116/

https://www.sciencedirect.com/topics/agricultural-and-biological-sciences/curio

Sajan, J. et al. 2012. An assessment of legally protected marine fauna in curio trade – A market study from Tamil Nadu, India. *International Journal of Conservation Science* 3(3): 217–230.

93. **These shark teeth/dried shark jaws are awesome – they also prove that the fishers aren't wasting any part of the animal.**
    - Which of your organs do you think aliens would like to take back home as souvenirs?
    - As in the case of many hunted/poached animals, the perpetrators often take only what is most valuable to them and discard the remains. That can range from the ivory of elephants, the horns of rhinos, the shells of sea turtles, the fins of sharks, to the swimbladders of certain fish (see entry Nr. 30). And remember, although poaching may sometimes be

motivated by poverty and hunger ("bush meat"), large-scale trafficking is motivated by pure greed.

- More than one-third of all sharks, rays, and chimaeras are now at risk of extinction because of overfishing, according to a new study re-assessing their IUCN Red List of Threatened Species extinction risk status. An estimated 100,000,000 (that's 100 million!) are killed every year. That's unsustainable. Taking any part of an animal home is certainly not an environmentally friendly or sustainable aspect of tourism and more often than not illegal as well. And, of course, it also helps to perpetuate the fishing effort for sharks and the market for these and other such dubious curio objects. Resist the temptation. One potential alternative is fossil shark teeth, which can reach truly impressive sizes far beyond those of any currently living species.

Grey, M. et al. 2005. Magnitude and trends of marine fish curio imports to the USA. *Oryx* 39(4): 413–420. https://doi.org/10.1017/S0030605305000967

Worm, B. et al. 2013. Global catches, exploitation rates, and rebuilding options for sharks. *Marine Policy* 40: 194–204. https://doi.org/10.1016/j.marpol.2012.12.034

94. **This dried seahorse looks so cute glued onto the driftwood – how could I resist?**
    - What other dried or stuffed animals do you have on display at home?
    - Don't tell me you have your poodles stuffed when they die.
    - Many of the world's 50 or so seahorse species are under unprecedented levels of threat and listed as protected species by the Convention on International Trade in Endangered Species (CITES). Beyond the degradation of their habitats (seagrasses, mangroves, coral reefs), a major threat is posed by the global trade in dried seahorses for traditional Chinese medicine: the estimated numbers vary from 20 million to 150 million specimens that are killed to "cure" asthma, arthritis, skin infections, baldness, infertility, impotence . . . claims for which there is little or no scientific evidence! Another few million are sold for the aquarium and curio trade. This calls for conservation efforts and awareness-raising. One promising approach is viewing them as a flagship species (species – often "charismatic" ones – that can be used as symbols in conservation campaigns to raise awareness and funding for specific conservation issues). Every dried seahorse is one less fulfilling its role in the environment and helping guarantee the survival of the population. Should a dead animal be the memento of your beach vacation? You might consider pestering the dealer, not being pacified by

claims of it being captive bred, and asking for proof of origin (and then walking away in disgust).

Cohen, F.P.A. et al. 2016. Seahorse aquaculture, biology and conservation: Knowledge gaps and research opportunities. *Reviews in Fisheries Science & Aquaculture* 25(1): 100–111. https://doi.org/10.1080/23308249.2016.1237469

Foster, S. et al. 2016. Opportunities and challenges for analysis of wildlife trade using CITES data – Seahorses as a case study. *Aquatic Conservation: Maine and Freshwater Ecosystems* 26: 154–172. https://doi.org/10.1002/aqc.2493

https://www.theseahorsetrust.org

95. **Shark finning may look brutal, but they are pretty brutal predators themselves, right? (and they don't really feel pain like we do) (Figure 18).**

Figure 18 Most of the millions of such finned sharks are finned while still alive – a horrendously inhumane and wasteful practice. To add insult to injury, many of the fins stem from endangered species.

- What other organisms would you cut up alive for profit?
- Shark finning is certainly among one of the most wasteful, cruel, and perverse uses of wildlife. One of the most universal tenets in hunting (not to mention animal farming) is to kill the animal as quickly and humanely as possible. Finning, however, often involves cutting the fins off the living, hooked animal and then throwing the still-living fish back into the ocean to a merciless death: The practice is also a major

waste of a valuable food source (the fins are worth an order of magnitude more than the meat). Somewhere between 30 million and 70 million sharks are traded across the global shark fin markets in Asia every year. The product itself is a luxury item and plays no role in fighting hunger or food security. Laws in many countries (like the Shark Finning Prohibition Act in the US and a similar one in the EU: EU Regulation No 605/2013) prohibit the practice and ensure that sharks are landed with their fins naturally attached. Theoretically. Nonetheless, the US and many countries continue to trade fins with countries with no finning laws. Globally, over 90% of shark catches are biologically unsustainable, with finning being a major driver of the problem. In the US, only 17% of the exploited shark stocks are sustainably fished: The country could preserve its key leading role in global conservation by enacting a shark fin trade ban, probably triggering similar legislative processes in other developed countries. Globally, the Marine Stewardship Council (MSC) certification assesses "naturally attached fins" as the "minimum acceptable" requirement.

As far as sharks being "brutal killers," all predators are emblematic of a certain "eat-or-be-eaten" fierceness of nature. Nonetheless, it is important to distinguish between malicious intentions and feeding-related intentions or necessities. The former could be attributed to humans as the most sentient of beings, the latter to sharks and, indeed, all predators. A predator's need to feed should never be taken as an excuse to abuse or kill it, much less unnecessarily reduce their populations to a minimum. And don't forget, even the act of herbivores partially or entirely eating plants has ecological repercussions that some folks would like to put a stop to (see Entry Nr. 124 regarding pain).

Ferretti, F. et al. 2020. Shark fin trade bans and sustainable shark fisheries. *Conservation Letters* 13(3): e12708. https://doi.org/10.1111/conl.12708

Gutteridge, A.N. et al. 2024. How are appropriate performance levels developed for MSC certification? A case study assessing shark finning. *Marine Policy* 163: 106119. https://doi.org/10.1016/j.marpol.2024.106119

http://www.sharkproject.org

Tønnessen, M. 2016. The semiotics of predation and the umwelten of large predators. In: Maran, T. et al. (eds.), *Animal Umwelten in a Changing World: Zoosemiotic Perspectives* (Vol. 18, pp. 152–183). Tartu Semiotics Library. University of Tartu Press.

96. **Eating sea turtle eggs can make me a better partner in bed. That's definitely a good cost-benefit proposition (Figure 19).**

Figure 19 We should enjoy the bounty that the sea provides. One exception might be sea turtle eggs, which are typically consumed less to still hunger than to support a misguided and misinformed machoism.

- So, your sex is more important than an intact environment for everybody?
- Exploitation and overexploitation of wildlife are conducted for many reasons, some involving human survival and food security, others more debatable such as recreational hunting. Some reasoning is downright frivolous and unsubstantiated, such as consuming wildlife or parts thereof as an alleged aphrodisiac, that is, to increase sex drive, sexual desire, and sexual performance. The latter is often the rumored benefit of consuming sea turtle eggs. In some areas of Mexico, for example, men consume the eggs of olive ridley turtles, which are on the brink of extinction, to combat erectile dysfunction. Many who consume these eggs exhibit traits associated with machismo, a form of masculinity

often characterized by objectification of women and hypersexuality, both of which drive men to seek aphrodisiacs. Importantly, eating sea turtles and their eggs is associated with multiple human health hazards (bacteria, parasites, biotoxins, environmental contaminants), with documented fatalities. Every egg removed from a freshly dug nest is one less hatchling that can help maintain the sea turtle population or, as is more often the case today, help to restore a severely depleted sea turtle population back to a healthy, sustainable size.

Aguirre, A.A. et al. 2006. Hazards associated with the consumption of sea turtle meat and eggs: A review for health care workers and the general public. *EcoHealth* 3: 141–153. https://doi.org/10.1007/s10393-006-0032-x

Harguth, B.M. 2019. *Machismo and Consequential Aphrodisiac Consumption in Oaxaca, Mexico*. Honors Thesis. 115. https://red.library.usd.edu/honors-thesis/115

97. **If you get rid of the sharks, all the animals they eat would survive, and the reef would have many more colorful fish (and the reefs would be safer for us too) (Figure 20).**

Figure 20 Shark feeding, here conducted as a group experience, reduces the natural environment into a poorly controllable event backdrop. Such sensationalism can never be sold as having conservation value. And if you are bitten, it may be little consolation that the Shark Accident Victim Network (SAVN) will support you in dealing with your trauma. (https://www.sharkvictimnetwork.org/) (Image from Shutterstock).

- You mean let's play "pimp the food web" and see what happens!
- Humans have always intervened in food webs. For example, by excessively exploiting a particular component. The repercussions were never really to our liking. Moreover, doing so to achieve some goal would require a very good understanding of the food web and its many interconnections, something we still lack even for the simplest of ecosystems. This is also reflected in our shaky attempts to mathematically model partial, not to mention entire, ecosystems.

  Maybe if we removed all the grey reef sharks, for example, they would be replaced by even more fearsome top predators, such as tiger sharks, great hammerhead sharks, or great white sharks. Oops. Maybe if we removed the reef sharks, their prey (smaller predators) would explode and eat all the many colorful small fishes (for example herbivores) in the reef that sharks cannot access. That could change the reef forever due to massive algal cover because the grazing fish are gone. Oops again. Or, alternatively, other colorful little fish would explode and nibble away at the reef and damage it in other ways. The fact is that we simply have no clue. Do you want to start that experiment? Importantly, many people (scuba divers) pay lots of money to see (and unfortunately to "swim-with" and feed) sharks. That type of income exceeds the sale value of shark meat manyfold – one strong argument for not killing sharks. And, as far as human safety is concerned, keep in mind that some of the largest shark species are harmless plankton feeders (e.g., whale shark, basking shark)! One often cited statistic is a maximum of 10 shark-related human deaths per year, whereas somewhere around 100 million sharks are killed by humans every year. The biggest killer in the water is the human being, not the shark.

  https://www.floridamuseum.ufl.edu/shark-attacks/yearly-worldwide-summary/

  Roff, G. et al. 2016. The ecological role of sharks on coral reefs. *Trends in Ecology and Evolution* 31(5): 395–407. https://doi.org/10.1016/j.tree.2016.02.014

98. **If this beach is destroyed, there are plenty of others for the sea turtles to lay their eggs in.**
    - How would you like moving from one home to another because someone keeps burning them down?
    - Like many reasonable-sounding arguments by non-experts about ecosystems and the environment, this one is flawed. Firstly, many beaches are widely separated by rocky shores or other coastlines unsuitable as nesting sites. Secondly, sea turtles first often return, sometimes two decades later, to or close to the beaches on which they hatched, a phenomenon known as natal homing and potentially involving some type

of imprinting. This can complicate nesting efforts if that and adjacent beaches have eroded away or been totally requisitioned for tourist facilities and other commercial interests. It is only a slight exaggeration to say that each major beach or stretch of sandy coastline is home to a subset of a particular sea turtle species, and losing those stretches can be regarded as a disturbance to that subpopulation. All seven species of sea turtles and many of their subpopulations are assessed in the International Union for Conservation of Nature's (IUCN's) Red Lists. With a few exceptions (categories: data deficient, least concern), the sea turtle listings are vulnerable, near threatened, endangered, or critically endangered. This calls for applying the precautionary principle and doing everything possible to ensure that all turtle nesting beaches (as one of their two habitats, namely beaches and the open ocean) continue to remain accessible.

In an enlarged context, why is the beach or parts thereof being damaged, despoiled, or narrowed? The chances are good that it is for commercial gain, for example, to construct luxury beach homes, a hotel complex, or beach bar with all their amenities from sunbeds and beach umbrellas to staked-off sand volleyball courts and jet-ski rental facilities. And you can be sure that you yourself will not be reaping any of the profits!

Lohmann, K.J. et al. 2013. Chapter 3: Natal homing and imprinting in sea turtles. In: Wyneken et al. (eds.), *The Biology of Sea Turtles* (Vol. III, pp. 59–78). CRC Press.

99. **If dolphins are so smart, why don't they simply leap over the fishing nets instead of being killed?**
    - Being smart isn't much help if you are victimized by people who are much smarter.
    - The intelligence of animals is a hot research topic. The simple general conclusion is that all have the smarts to survive in their respective environments (otherwise the species wouldn't be there). That's also the basic answer to the debate about which is smarter, dogs or cats. In this vein, the often-heard remark that a particular animal is optimally adapted to its environment is an absolutely self-evident and needless statement. A second conclusion, confirmed time and time again by animal behavior experts and cognition researchers, is that many animals are much smarter than we thought. When we alter their environments, however, they are forced to face tasks and hurdles for which they did not evolve. In the case of open-water cetaceans such as whales and many dolphins, the oceans have always been a barrier-free environment, and almost everything is worth nibbling at as potential food. This means that their

cognitive and behavioral repertoire is not adapted to reacting to virtually invisible fishing nets or lengthy roping in the water. This explains why millions of dolphins have been trapped and killed in nets used in tuna fishing, for example, and why cetacean entanglement in and ingestion of plastic items has become one of the greatest threats to their survival. Over 300,000 whales and dolphins die annually due to entanglement in active fishing gear, with additional unknown numbers also entangled and killed by and in marine debris.

https://iwc.int/management-and-conservation/entanglement

Simmonds, M.P. 2006. Into the brains of whales. *Applied Animal Behaviour Science* 100(1–2): 103–116. https://doi.org/10.1016/j.applanim.2006.04.015

Würsig, B. 2009. Intelligence and cognition. In: Perrin, W. et al. (eds.), *Encyclopedia of Marine Mammals* (2nd ed., pp. 613–623). Academic Press. https://doi.org/10.1016/B978-0-12-373553-9.00143-7

100. **We only took a few truckloads of sand from the sea turtle nesting beach for the construction project – it hardly made a dent (Figure 21).**

Figure 21 Sand removal from beaches is one of the quickest and most perfidious way to destroy what may be one of the only "natural" ecosystems that most people will encounter these days. This is doubly disgraceful here because it's on a sea turtle nesting beach.

- And how long do you think it takes for nature to replace all those sand grains?
- And when do you and others plan to come back for the next construction project?
- Sand beaches, next to the open ocean, are the two crucial habitats for sea turtles. And sand beaches around the world are threatened by erosion. This, for example, reflects construction on and behind beaches along with damage to sand dunes and mangrove forests. Increased storm activity and rising sea levels due to climate change also play a role. All structural attempts to save beaches either are basically failures or merely shift the problem to the next section of the beach. And replenishing the sand on beaches with "offshore" sand is both very expensive and devastates the sand-dwelling invertebrates from both source and destination. Moreover, such "beach nourishment" efforts can alter the sand grain size composition of the beach and its microclimatic conditions (e.g., for sea turtle nests). Accordingly, any sand removals should be avoided and severely prosecuted. In some countries, it is even forbidden (and punishable) for visitors to take a handful of "souvenir" sand back home. Yes, "natural" souvenirs may seem unique and genuine (and bring back good memories that cannot be bought in stores). Nonetheless, all of them – including sand, shells, and other curio items, whether collected or purchased – remove components from either your vacation environment or some faraway marine ecosystem. Reduce your footprint and leave them where they belong. Report companies that send their trucks to steal sand from the beach to the authorities. Take a photo as documentation (as surreptitiously as possible). A camera can be a mighty weapon!

https://green-travel-blog.com/green_pearls_insighs/sand-as-a-travel-souvenir-why-its-better-to-leave-it-behind/

Pabian, A. et al. 2020. Young people collecting natural souvenirs: A perspective of sustainability and marketing. *Sustainability* 12(2): 514. https://doi.org/10.3390/su12020514

Rangel-Buitrago, N. et al. 2023. The global impact of sand mining on beaches and dunes. *Ocean & Coastal Management* 235: 106492. https://doi.org/10.1016/j.ocecoaman.2023.106492

Wooldridge, T. et al. 2016. Effects of beach replenishment on intertidal invertebrates: A 15-month, eight beach study. *Estuarine, Coastal and Shelf Science* 175: 24–33. https://doi.org/10.1016/j.ecss.2016.03.018

101. **As a biologist you can't have anything against trees, so why shouldn't we be planting trees on this sea turtle nesting beach?**

- If trees could grow on the beach naturally, they would already be there.
- Beaches are natural ecosystems in their own right, with a specialized fauna and flora distributed in well-recognized zones. Larger, longer-lived tree species are typically unable to withstand the rigorous conditions on beaches (wind, moving substrate, waves, salt) and are therefore first found further on the landward side of beaches, often behind the sand dunes if these are present. If trees grew naturally on the beach in question, they would be there without human intervention. Planting trees directly on the beach is therefore ecologically questionable and rarely successful (and might involve planting "hardier" non-native species, which is always a problem). Importantly, the main reason for such plantings is often to accommodate tourists (shade) rather than any environmental considerations. On sea turtle nesting beaches, trees should not be planted because:
  - their trunks are a barrier for the turtles;
  - their roots can extend far and wide and be dense, preventing the adult females from digging into the sand and the hatchlings from emerging should the female have nonetheless been successful;
  - the tree's shade lowers the sand temperature and therefore the incubation temperature, which can prolong embryonic development, kill the embryos, or alter the sex ratio of the hatchlings (which is determined by the temperature); and
  - trees, as magnets for visitors, tend to be beach litter hotspots.

de Vos, D. et al. 2019. Effect of introduced Casuarina trees on the vulnerability of sea turtle nesting beaches to erosion. *Estuarine, Coastal and Shelf Science* 223: 147–158. https://doi.org/10.1016/j.ecss.2019.03.015

Wood, A. et al. 2014. Sun exposure, nest temperature and loggerhead turtle hatchlings: Implications for beach shading management strategies at sea turtle rookeries. *Journal of Experimental Marine Biology and Ecology* 451: 105–114. https://doi.org/10.1016/j.jembe.2013.11.005.

102. **I only broke a small piece of coral off an enormous colony as a souvenir – surely that can't hurt!**
    - So, it's OK for each of the 100 divers here today to take a piece with them?
    - So, it's OK if I cut just a teeny-weeny bit off your little finger?
    - Coral breakage, whether by divers and snorkelers or due to storms, poses a major threat to the reefs. Corals grow very slowly, and the piece you broke off accidentally may represent years or even decades

of growth. If you broke it off on purpose, then you will no doubt have done so wearing gloves – and therefore have understood why divers are generally not permitted to wear gloves in coral reefs. Breakage is a severe injury, and the coral must repair the damaged remaining parts. Those injuries are also susceptible to infection and disease, which have the potential to kill the entire colony. Finally, diseases may spread from colony to colony. Taking a piece of broken-off coral home is perhaps the worst of all reef-related souvenirs – and could make for a very unpleasant and expensive confrontation at airport customs. This also holds true for weathered, dead pieces (coral rubble), whether found underwater or on the beach, and for fragments purchased at souvenir or curio shops, whether they be single pieces or part of some iffy sculpture.

Brandt, M.E. et al. 2013. Disturbance driven colony fragmentation as a driver of a coral disease outbreak. *PLoS One* 8(2): e57164.

Lamb, J.B. et al. 2014. Scuba diving damage and intensity of tourist activities increases coral disease prevalence. *Biological Conservation* 178: 88–96. https://doi.org/10.1016/j.biocon.2014.06.027

103. **If motorboats pose such a threat to whales and dolphins due to collisions, propellers, and noise, then why don't all whale watching operations use sailboats?**

- Are you as a pedestrian more likely to be hit unaware by a silent electric vehicle or a noisier one?
- Many "platforms" are available for whale watching, including motorboats, sailboats, cruise ships, inflatables, kayaks, helicopters, and airplanes. Using sailboats as whale watching vessels seems like a logical and ecofriendly idea. Nonetheless, experts do not necessarily recommend it because, in this case, the noise produced by motorized vessels gives the cetaceans a heads-up that something has entered their area and that they should be aware of it. Some species may change their behavior or take evasive action from the onset. Silent vessels such as sailboats fail to give such advance warning. One solution when using sailboats is to keep the motor running (in neutral). From the practical perspective, dolphins and porpoises ("small cetaceans") can be unpredictable in their movements, and sailboats lack the maneuverability and aren't really capable of getting to the site quickly enough or then keeping up with the animals. This makes sailboats more of "a great day out on the water" than a commercially viable whale watching operation. Note also that offshore sailboat racing has now been recognized as a (collision) threat to whales. Of course, motorized

boats also pose problems of their own. The risk of collisions is highest during higher-speed trips to and from the whale watching site itself, as is the presence of multiple boats maneuvering around a cetacean group. The latter situation may be illegal, detracts from the visitor experience, can lead to negative reviews of the whale watching trip on social media, and has raised the issue of needing to certify sustainable boating operators with an ecolabel.

Finkler, W., Higham, J. 2004. The human dimensions of whale watching: An analysis based on viewing platforms. *Human Dimensions of Wildlife* 9(2): 103–117. https://doi.org/10.1080/10871200490441757

Ritter, F. 2012. Collisions of sailing vessels with cetaceans worldwide: First insights into a seemingly growing problem. *Journal of Cetacean Research and Management* 12(1): 119–127.

104. These plastic pieces are so small that they can't possibly harm any marine life (Figure 22).

Figure 22 The closer you look, the more you'll find. The plastic industry pours these pre-made pellets into the molds that make virtually all plastic items. These "primary microplastics" are distinguished from the tiny fragments (secondary microplastics) formed by decomposing litter on beaches and out at sea.

- The smaller they are, the more likely you are to breathe them in and consume them with your (sea)food.
- Viruses and bacteria are pretty small, too.

- Based on size, researchers broadly distinguish between mega-plastics (10 cm to several meters), macro-plastics (larger than 0.5 or 1 cm), microplastics (1 μm–5 mm), and nanoplastics (below 1 μm). Microplastics are further divided into primary microplastics (produced at that size) and secondary microplastics (formed by the decomposition of larger items). The former includes the pellets that the plastic industry produces as raw material for the moulds that form all other plastic items, or the microbeads in personal care items. Millions of tons of pellets are produced, and enormous amounts are lost at sea or enter the oceans through rivers. Secondary microplastics can comprise small fragments and fibers, that is, the breakdown products of larger pieces. Simply put, all categories are major threats to marine life. The largest items entangle marine life. Smaller ones are deliberately (but erroneously) ingested and may either pass through the digestive tract or block it, leading to slow starvation. The smallest are consumed unintentionally and unaware and can end up in the internal organs and cells of everything from zooplankton to whales. Micro- and nanoplastics are therefore among the hottest items in marine pollution research today. They are also the most likely to enter the human body via food or breathing and are increasingly being recognized as a human health hazard.

Sharma, S. et al. 2024. Understanding microplastic pollution of marine ecosystem: A review. *Environmental Science and Pollution Research* 31(29): 41402–41445. https://doi.org/10.1007/s11356-023-28314-1

Sina Pourebrahimi, S., Pirooz, M. 2023. Microplastic pollution in the marine environment: A review. *Journal of Hazardous Materials Advances* 10: 100327. https://doi.org/10.1016/j.hazadv.2023.100327

105. **This jewelry/these items made out of sea turtle shells are natural products and help reduce plastic use.**
   - Heroin and cocaine are also natural products, but that doesn't make them good.
   - I'm sure that turtle gladly gave up its shell so you could groom your hair with a "natural" comb.
   - In the early days, almost all items used by humans were handcrafted from natural materials. This ranged from stone arrowheads or antler buttons to umbrella ribs and corset boning made of whalebone. Today, wildlife no longer needs to be killed for any such product, and plastics are a justified replacement material. On the luxury market, certain objects continue to be offered made from the plates (scutes) of the hawksbill sea turtle. This often runs under the heading of "tortoiseshell" or, in Japan, the bekko industry. The items range from

necklaces, brooches, and cufflinks, to shoehorns, model sailing ships, cigarette holders and cases, or key rings. In all cases, the sea turtles were killed to obtain this material, typically without further using the animal as food (akin to the ivory of elephants or the horns of rhinos). Whether in a jewelry, curio, or high-end accessories shop, whether new or antique, online or not, you should avoid buying such items. Consumer purchasing choices can make a difference. Use your wallet to help put them out of business. That might also save you considerable time and money when going through customs on your way home.

Canin, J. 1991. International trade aspects of the Japanese Hawksbill Shell ('Bekko') industry. *Marine Turtle Newsletter* 54: 17–21. http://www.seaturtle.org/mtn/archives/mtn54/mtn54p17.shtml

LaCasella, E.L. et al. 2021. Mitochondrial DNA profiling to combat the illegal trade in tortoiseshell products. *Frontiers in Marine Science* 7: 595853. https://doi.org/10.3389/fmars.2020.595853

Nahill, B. 2021. Trade in Hawksbill Shell Still Active Globally. SWOT Report 16. https://www.seaturtlestatus.org/articles/swot-report-16-trade-in-hawksbill-shell-still-active-globally.

106. **I can't afford to use all-new ecofriendly fish hooks/nets/equipment for my fishing operations – your organization will have to pay for it.**

- You mean someone else has to pay for you to do the right thing?
- Very often, new, alternative gears are available to replace destructive fishing techniques. This can involve whale-friendly lobster traps (the lines between the traditional pots and those leading to the surface can ensnare or entangle whales). Another example is replacing the use of chemicals to extract colorful reef fishes for the aquarium trade with hand nets. The classic example is replacing the traditional J-shaped fishing hooks with circle hooks (so-called C-hooks). The former have been perfected over centuries to catch fish, but they also snare sea turtles, seabirds, sharks, and many other species that represent unwanted bycatch. The latter have been designed to catch fish too, but do not so easily get hooked in the mouths of sea turtles. Introducing new, ecofriendlier gear faces two hurdles. The first is that fishers are often initially skeptical about the performance of the new gear. Will it catch the same number of fish as the old system? This takes an effort on the part of fisheries or conservation managers, backed up by scientific research and practical demonstrations. One line of argumentation is that the new system boosts the survival rate of bycaught animals (the desired conservation outcome) and reduces overall bycatch, which is often time-consuming and dangerous to remove from the lines. The second hurdle is cost. All fishing

gear is designed to take severe punishment and therefore built to last (i.e., expensive); replacing a full set can be costly. In the case of longline fishing, for example, a fisher might have thousands of hooks on one line. Replacing them all with C-hooks is an expensive proposition. In some cases, the respective programs can be funded, organized, and implemented by international and national NGOs, regional fisheries management organizations, national fisheries agencies, or fisher cooperatives. In most cases, it's ultimately a win-win situation for both the fishers and conservation managers.

Andraka, S. et al. 2013. Circle hooks: Developing better fishing practices in the artisanal longline fisheries of the Eastern Pacific Ocean. *Biological Conservation* 160: 214–224. https://doi.org/10.1016/j.biocon.2013.01.019

Pacheco, J.C. 2011. A comparison of circle hook and J hook performance in a western equatorial Atlantic Ocean pelagic longline fishery. *Fisheries Research* 107(1–3): 39–45. https://doi.org/10.1016/j.fishres.2010.10.003

Virgili, M. et al. 2024. Engaging fishers in sea turtle conservation in the Mediterranean Sea. *Marine Policy* 160: 105981. https://doi.org/10.1016/j.marpol.2023.105981

107. Why worry about the garbage on the shore – we have an annual beach clean-up anyways (Figure 23).

Figure 23 Marine debris or beach litter has many sources, including boats, rivers and streams, and beachgoers. What are the chances that every last item here will be taken back home?

- So, I can dump garbage at your doorstep because it will be picked up by volunteers sometime this or next year?
- "Never worry about your heart until it stops beating." (Elwyn Brooks (E.B.) White)
- We are literally surrounded by litter (space debris/space junk – hundreds of millions of items ranging from nuts and bolts to entire defunct satellites circling overhead). And we tend to oversee or ignore the untold items littering our streets. But litter on our vacation beaches perplexes and angers us. That makes beach litter or marine debris one of the hottest issues in marine research these days. Much, but by no means all of it, revolves around plastic. The annual International Coastal Cleanup (ICC), organized by the Ocean Conservancy in partnership with volunteer organizations and individuals around the globe, is an amazing effort to highlight the problem and take direct action. Moreover, it is apparently the largest single-day volunteer activity in the world (in September, about 1 million helpers every year). Unfortunately, the garbage can't be removed from every beach and shoreline, and it accumulates faster than it can be removed. It is often piled deepest on islands that are uninhabited (and never cleaned). Researchers can use these shores as the benchmarks for actual accumulation rates and densities. Some beachside hotels now even offer beach-cleaning sessions as an activity for their guests. Each participant gets a large garbage bag and heads on down (or up) the beach. Tellingly, some don't even get out of eyesight before their bags are filled to the brim. Importantly, it's not merely an aesthetic problem: most items are harmful to marine life, and some pose a threat to beach-goers as well (nails in wood, broken glass, dog feces, syringes, sharp, rusted metal, improperly discarded medical wastes). Remember, beaches are a barefoot environment! Why not get active and sign up for a cleanup or even start one of your own in your area? You could also just simply take one more piece of litter back home than you brought with you.

  https://oceanconservancy.org/trash-free-seas/international-coastal-cleanup/

  Stachowitsch, M. 2019. *The Beachcomber's Guide to Marine Debris*. Cham, Switzerland: Springer. 358 pp.

108. **The whale populations are growing and eating all our fish, so we have to start culling them.**
   - So, you want to kill all birds because some eat the cherries from your tree?

- Unfortunately, regulating any food web by killing off a particular group of organisms has never been successful.
- Which whale species do you want to start with?
- This line of thought was advanced by the Japanese delegation at the International Whaling Commission (IWC) – and also used as an argument to support lethal scientific whaling (e.g., to examine stomach contents and see what they were really eating). Over the years, this category of whaling received such strong pushback from most cetacean scientists that Japan withdrew from the organization in 2019. Properly addressing the whales-eat-fish issue means going beyond the pure "gut content" approach. It requires a better understanding of whale distributions, prey distributions, prey availability, and prey choices at different seasons for various migratory whale species. Such information is only partially available.

In a simplified food chain scenario, if whales directly ate the same species we do, and if the whale population increased, then they might, at certain sites and certain times of the year, reduce that fish population. Conversely, if the whales also consumed the predators of some of the fish that we eat, and the whale population increased, then they might reduce the populations of those predators and allow the fish that we eat to increase. In line with the saying "If you can't solve the problem, enlarge the context," the question needed to be expanded. What cetaceans actually consumed more fish? Were they certain large baleen whales (some of which were targeted in scientific whaling) or were they rather the toothed whales (dolphins and porpoises)? If the latter, as top predators, consumed too many fish, then shouldn't we be culling them instead? Hey, do the baleen whales actually eat mostly fish? I thought they mostly ate planktonic organisms such as krill, in which case they wouldn't necessarily be competing with us for fish. (And while we are on the subject, strict quotas are now necessary for krill because we have finally figured out how to exploit them commercially.)

Going even one step further outside the "marine ecology" box, are people starving because they are not getting enough fish? If there are not enough fish, who or what is actually responsible for that? Might it be our overfishing? Do whales pose a food security risk because they eat "our" fish. Are whales a food security issue because of the current moratorium on commercial whaling, and that "missing" whale meat would be a big help? These and other considerations (see also entry Nr. 3) explain why it took years of effort to counter the "whales are eating our fish" point of view and why certain aspects (food security) continue to be contentious.

Ruzicka, J.J. et al. 2013. Dividing up the pie: Whales, fish, and humans as competitors. *Progress in Oceanography* 16: 207–219. https://doi.org/10.1016/j.pocean.2013.07.009

109. **OK, so navy sonar is killing dolphins, but our national security is much more important than a few dolphins.**

- You mean you feel safer when dolphins are killed?
- I've heard of national security being an excuse for taking action against virtually everything . . . but *killing dolphins*?
- That sounds like a good excuse for any kind of collateral damage by the military.
- The general understanding today is that environmental degradation poses a greater threat to cetaceans (whales, dolphins, porpoises) than current whaling/hunting quotas. Current quotas, thanks to the International Whaling Commission (IWC), are very modest compared with the slaughter of past centuries. Noise is one of the many such recognized environmental concerns (including chemical pollution, collisions with ships, bycatch, marine debris, habitat degradation, and climate change). The noise threat is severe considering that most cetaceans communicate using sound (experts refer to the whales' "soundscape"). The oceans have become an increasingly loud environment that reduces the communication space for cetaceans, and sound at very high levels can damage hearing capabilities and even kill the animals. Military (navy) sonar – low- and mid-frequency active sonar (LFAS, MFAS) – is used to detect submarines and is one such potentially lethal sound source. It can cause temporary (TTS) and permanent threshold shifts (PTS) in the hearing of cetaceans but can also severely damage their hearing organs or cause them to surface too quickly and die. Deep-diving species are particularly affected, particularly beaked whales, but also pilot whales, minke whales, several dolphin species, and harbor porpoises. Numerous strandings around the world have been coincident with naval exercises (and have necessitated coining a new technical term – unusual mass strandings). This has led to injunctions against military exercises using sonar, resolutions from international treaty organizations, and statements of concern by professional organizations. It also raises the question whether the (US) Navy should be exempt from environmental regulations (which forbid harassing or killing marine mammals) during peacetime. Many militaries have committed to investigating and mitigating their activities to protect marine mammals, but this also requires committing to conducting adequate baseline monitoring in areas where exercises routinely occur. And, at minimum, it requires having

observers in the area to ensure that cetaceans are not present during "war games" and other exercises.

Parsons, E.C.M. 2017. Impacts of navy sonar on whales and dolphins: Now beyond a smoking gun? *Frontiers of Marine Science* 4: 295. https://doi.org/10.3389/fmars.2017.00295

110. **I've seen fish and invertebrates living in and on bottles and other garbage on the bottom, so it's actually helping them.**
    - Well, people living in cartons and among other garbage are often referred to as being destitute or homeless, right?
    - Many sessile marine invertebrates do settle on any available hard structures because the sand and mud bottoms that characterize most of the world's oceans do not provide a solid surface to attach to. Smaller fish and mobile invertebrates may also seek shelter in litter on the seafloor. This, however, should not be interpreted as being a "positive" effect of marine debris or litter. Bottom-living organisms may have little other choice these days. In many shallower waters, the typical physical structures that normally provide a place for such organisms to settle (shells, invertebrate aggregations, tubeworms, etc.) have often been crushed and fragmented by destructive commercial fishing gear such as bottom trawls and dredges. (This will also become an issue if deep-sea mining for nodules is approved.) Importantly, much of the litter gradually decomposes, so that organisms that settle on or in it may have only temporary housing. This issue is related to that of artificial reefs (see entry Nr. 60) or to lost, abandoned, or discarded fishing gear such as "ghost pots" that keep on capturing and killing crustaceans (lobsters, crabs) that fishers will never harvest. Other forms of litter can entangle and kill bottom-dwelling animals. Finally, marine debris ultimately fragments into ever-smaller particles, leading to the microplastics and nanoplastics problem that affects both marine life and human consumers of seafood.

    Angiolillo, M. et al. 2021. Distribution of seafloor litter and its interaction with benthic organisms in deep waters of the Ligurian Sea (Northwestern Mediterranean). *Science of the Total Environment* 788: 147745. https://doi.org/10.1016/j.scitotenv.2021.147745

111. **I don't see the purpose of a "no-wake" zone because storms make much bigger waves than my boat.**
    - Isn't "smooth sailing" in life better than always "making waves"?
    - No-wake zones can be prescribed for many reasons. They may help reduce the erosion of river banks and other shorelines by countless

passing motorboats at speed. Waves higher than ca. 30 cm (12 inches) erode vegetated shores or compromise marsh survival. These no-wake zones also reduce the disturbance of birds and other animals inhabiting the reeds at water's edge and, more generally, reduce the sediments that are resuspended by propellers and cloud the water. Above all, slower speeds increase the chances that captains will not oversee organisms resting on the surface or coming up to breathe (turtles, manatees, dolphins). Slower speeds also reduce the threat and severity of injuries to wildlife if collisions do occur. Recreational boating means being out on the water and enjoying the surroundings. High-speed travel won't bring that pleasure. Finally, no-wake zones are typically limited to more sheltered areas that are not subject to constant wind-induced wave action. Enjoy that zone and wait until you are out in the open water before opening up the throttle. That was the carrot argument. The stick arguments are heavy fines and, if that proves unsuccessful in reducing speeds, then prohibitions on boats of a certain size or engine size, or limiting the number of passages, are up for discussion.

Bilkovic, D.M. et al. 2019. Defining boat wake impacts on shoreline stability toward management and policy solutions. *Ocean & Coastal Management* 182: 104945. https://doi.org/10.1016/j.ocecoaman.2019.104945

7
Hunting and Fishing

112. **Why can't we eat this (endangered) species? Aren't humans more important than animals?**
   - We can't eat ourselves out of any dilemma.
   - Define "more important."
   - So, you are arguing that the most important species for the environment is *Homo sapiens*?
   - Let's first ask why the species is endangered. It's probably because we have either destroyed its habitat or hunted/fished it to some pitifully low level (overexploitation). Continuing on that path bodes ill both for that species and for our own physical and psychological welfare in the future: harming our natural environment and decimating its inhabitants is the quickest way to harm ourselves. Overall, this hotly debated topic runs under the heading of "bush meat" or "wild meat," with an added dollop of "food security." It boils down to the question of how to save an exploited species while ensuring food security and the livelihoods of people. Some people who formerly relied on bush meat are ultimately forced to switch to alternate forms of protein and livelihoods when wild meat becomes rare. Most are insightful when this happens, that is, they recognize that their quarry has been reduced to such a level that it simply no longer pays in effort or time to hunt them (in commercial fisheries, for example, a status referred to as functional extinction). And speaking of "pay," some species become rare because hunters find it more profitable to kill the animals not to nourish themselves and their families but to sell them (or certain parts of them) on the market. Be that as it may, at that point the recovery of the species typically entails major and expensive management measures (and strict hunting bans). Ultimately, the answer lies in looking at the wider picture ("if you can't solve the problem, enlarge the context"): what kind of environment do we

DOI: 10.1201/9781003757474-7

(that includes the hunters, of course) want for ourselves and our children? Will we be happier and healthier in a functioning ecosystem with all its inhabitants or be satisfied with seeing "our" former wildlife in a cage or display case at a zoo or museum?

Rao, M., McGowan P.J.K. 2002. Wild-meat use, food security, livelihoods, and conservation. *Conservation Biology* 16(3): 580–583. http://www.jstor.org/stable/3061202

Milner-Gulland, E.J. et al. 2003. Wild meat: The bigger picture. *Trends in Ecology and Evolution* 18(7): 351–357.

113. **My family has fished here/hunted here for generations, why should we have to stop now?**

- Many families held slaves for generations. Stopping that was a good idea, wasn't it?
- If the animal population you hunt has remained stable over all those generations and you can guarantee future sustainable kills, then OK.
- Yes, people have always hunted/fished in their vicinity. Historically, however, this has involved simple hunting tools, access by non-mechanical means, and put much-needed food on the table. (Early-modern extravagant hunting by monarchs and other royalty is no doubt an exception.) Today, high-powered firearms and everything from skidoos to quads and powerful boats (not to mention aircraft) have considerably extended our sphere of action/damage and our hunting passions. Again, it's helpful to enlarge the context. Viewed more broadly then, human history is characterized by changes in our professions and ways of life. For example, many jobs and activities that were once crucial have gone out of existence due to new technologies. And each such change has apparently made us fitter – otherwise the human race wouldn't be so successful and approaching the 10 billion mark. Sometimes we simply have to acknowledge that some of our past ways – especially when amplified by modern technology – are counterproductive and no longer fit into today's world. Hunting could be viewed from that perspective. Is hunting necessary for your family's nutrition? Do you really need that trophy mounted in your den? Can illegal hunting and poaching be adequately controlled? Does the increasingly deteriorated and shrinking natural habitat in your area make it difficult enough for wildlife to survive – without additional hunting pressure?

Harrison, R.D. et al. 2016. Impacts of hunting on tropical forests in Southeast Asia. *Conservation Biology* 30(5): 972–981. https://doi.org/10.1111/cobi.1278

114. **Hunting this species is part of our culture, and it would threaten us as a people if we had to stop.**
   - Well, genital mutilation is also a part of many cultures . . .
   - So, you're arguing for continuing to hunt/fish until the bitter end (extirpation) of that species?
   - That's a powerful argument in a day and age when people around the globe are struggling to preserve their traditions and cultures – and the value of doing so has become universally acknowledged. Few people would question the need for hunting when that has been a historical keystone of an indigenous culture and also fulfils crucial nutritional needs. Unfortunately, any cultural practice that ultimately destroys the very basis on which that culture thrives or once thrived is heading for trouble. Some international organizations, such as the International Whaling Commission (IWC), for example, have given great consideration to this in continuing to issue whaling quotas for indigenous peoples. The IWC has also long attempted to incorporate and distinguish both cultural and nutritional needs, whereby the latter does not include broader economic considerations such as the sale of meat. This ultimately proved to be very contentious and burdensome in the framework of one of the three recognized categories of whaling, namely, aboriginal subsistence whaling. One complication is determining whether the needs of a growing human population will ultimately overburden what was traditionally extracted from the environment by a smaller, historical indigenous population. Are there alternative options for food? Ethical considerations were also much debated under the title "humane killing" because hunting with traditional weapons tended to involve long times to death. These and many other considerations underscore the problems involved in addressing this issue. Accordingly, the new system at the IWC requires only "a description of the hunt," with no "needs statements" required. Another way of relativizing the issue is to examine whether you feel your "overall" culture would truly be threatened without a hunting aspect, and whether you actually still use traditional weapons and continue to exercise and rely on the ancient and often elaborate cultural practices historically associated with the hunt.

     ASW03 Quantification of Subsistence and Cultural Need for Bowhead Whales by Alaska Eskimos (USA). 2012. https://iwc.int/document_2967

115. **We as indigenous people live in harmony with nature and understand our environment, so our hunting/fishing is sustainable.**
   - If everything is so harmonious, why are we talking extirpation/extinction?

- So, will you continue to hunt with the ancient weapons and the traditional trekking practices to ensure that the hunting remains sustainable?
- Research in fields such as archeology, historical ecology, and conservation paleobiology has shown that the extirpation or extinction of species and the loss of biodiversity have occurred time and time again despite humans claiming to have understood the ecosystems they inhabited. Their knowledge is often subsumed under the heading of traditional environmental knowledge (TEK). TEK is defined as "a cumulative body of knowledge, practice and belief evolving by adaptive processes and passed down through generations by cultural transmission, about the relationship of living beings (including humans) with one another and with their environment" (Berkes et al., 2000). It is site specific and represents the often verbally passed-on information necessary to reconcile both nature protection and cultural survival. As such, TEK can be considered as one potentially viable approach to "managing" the respective ecosystem.

Note, however, that earlier human populations were unlikely to have had some deeper understanding of the ecosystem than that available to the modern scientific community (but see entry Nr. 12). In some cases, they apparently did severely alter the environment or decimate the surrounding wildlife many centuries and even millennia ago. Accordingly, the demise of entire civilizations in relation to environmental degradation has become a matter of modern scholarly debate. In other cases, early hunts with their limited mobility and inefficient traditional hunting and fishing techniques clearly did less damage than subsequent commercial harvests. Modern descendants, however, will obviously use the latest weaponry and other modern equipment ("From Kayak to Yamaha"; p. 31, Hansen, 2002). That can lead to ecologically and economically damaging hunts. Of course, the problem is of global rather than mere regional interest because wildlife is part of humanity's common heritage.

That said, one strategy being discussed to protect species or entire landscapes is to more strongly involve indigenous peoples. Examples include giving local people a stake in the success of wildlife preserves (ranging from special reserves to marine sanctuaries from Africa to Palau). Some even envision handing land (back) to indigenous peoples for a potentially better stewardship ("indigenous futurism"). If valuable information on an environmental issue is inadequate or lacking (when neither TEK nor solid research results are available), then one strategy today is to invoke the precautionary principle. It basically states that if an action might pose a risk of harm to the public or environment, then that policy or action should not be carried out. This might also be applied to forbid or restrict certain hunts or hunting practices.

Berkes, F. et al. 2000. Rediscovery of traditional ecological knowledge as adaptive management. *Ecological Applications* 10(5): 1251–1262. https://doi.org/10.1890/1051-0761(2000)010[1251:ROTEKA]2.0.CO;2

Hansen. K. 2002. A farewell to Greenland's wildlife. Bæredygtighed. Originally published in Denmark, 2001, by Gads Forlag. 154 pp.

Rick, T.C., Erlandson, J.M. (eds.). 2008. *Human Impacts on Ancient Marine Ecosystems: A Global Perspective*. University of California Press. 319 pp.

116. **Officer, that animal I caught/shot/ran over by accident is already dead, so if I don't get to keep and eat it, it would be a waste of meat.**

- Let's define "accidental."
- You didn't know the gun was loaded?!
- That's a tricky question. On the one hand, nothing should be wasted these days. On the other hand, "rewarding" a kill that resulted from an accident or that perhaps even involved illegal behavior might simply promote that undesired activity. That's one reason why road kills in most countries must be reported to the authorities, and the kills themselves cannot be taken home for barbecuing. That logic is also behind the repeated burning of the many tons of ivory from illegally killed elephants. The collected information from road kills can provide input into decisions regarding the management of those species that are particularly subject to such kills. It can also be used to improve traffic and road safety. Of course, shooting an animal by accident is a many-faceted hunting topic in and of itself. Any such event probably involves one or more of the following: alcohol, poor eyesight, poor training, poor knowledge of local wildlife species, trigger-happiness, or a "that-animal-doesn't-belong-here" mentality. None of the above should be rewarded with food on the table or a trophy on the wall. An insightful story illustrates this problem. A farmer in the US painted the word "COW" in large red letters on all his cows at the onset of the hunting season to avoid livestock losses. When interviewed and asked what he wrote on the bulls, he responded "I also paint 'COW' – it's no use confusing them city hunters."

A similar but larger-scale issue arose regarding using the meat of animals that were officially killed for food purposes but then used otherwise. In one such case, the meat of whales – killed under the aboriginal subsistence whaling category of the International Whaling Commission

(IWC) – was used not primarily as human food, but as feed for mink farms. That use was ultimately abolished. A broader case involved the whales taken under the so-called scientific whaling category ("Whaling under Special Permit") in the framework of the IWC. Here, in contrast to the accidental road kills discussed earlier, an alternative decision was made. These whales were ostensibly killed solely for scientific investigations (as opposed to commercial whaling and aboriginal subsistence whaling). Nonetheless, the thousands of tons of meat from those whales killed over the decades in this whaling category were allowed to be sold legally on the market. Any other solution would have involved major wastage. This, however, created unresolved problems in that better supervision and (genetic) controls would have had to be in place to prevent violations. Violations included (a) other, illegally killed whale species, (b) the approved species, but killed in non-authorized locations, (c) imports from abroad, (d) meat from dolphins and other mammals (horsemeat . . .) sold as whale meat – all under the guise of permitted whale species and authorized amounts. These issues, along with the high number of whales killed over the long term and the fact that most were being killed in a recognized whale sanctuary, hounded the whaling nation involved for years. It ultimately ceased scientific whaling and left the IWC.

Lukoschek, V. et al. 2009. High proportion of protected minke whales sold on Japanese markets is due to illegal, unreported or unregulated exploitation. *Animal Conservation* 12(5): 385–395. https://doi.org/10.1111/j.1469-1795.2009.00302.x

Sadleir, R.M., Linklater, W.L. 2016. Annual and seasonal patterns in wildlife road-kill and their relationship with traffic density. *New Zealand Journal of Zoology* 43(3): 275–291. https://doi.org/10.1080/03014223.2016.1155465

117. This fish is just a few inches below the legal limit – give me a break!

- Size matters.
- If a child is a few inches too short to take that killer rollercoaster ride, should it be allowed to go anyway?
- If you are just a few inches too tall to be a fighter pilot, should the Air Force take you anyway?
- The size of a particular fish makes a big difference in its many roles in the ecosystem and in its contribution to the survival of the population. Ecologically, older and larger individuals, for example, may take on the role of top predators, while the younger individuals are a rung or two lower on the food chain. With regard to reproduction,

older females produce larvae that have a better survival rate than do larvae of younger specimens. On the one hand, a few inches can decide whether a fish has had the chance to reproduce at all. On the other hand, larger individuals have an exponentially greater fecundity/fertility. For example, one 61 cm female red snapper produces the same number of eggs as 212 females at 42 cm each (a 19-cm or 7.5-inch size difference). This can lead to two opposing views on size limits. The first and most common is that smaller individuals should be released so that they have a chance to reproduce. Alternatively, large individuals should be released because they contribute the most to maintaining the population. Such considerations play a role in recreational fishing as well in regulating (or banning) recreational spearfishing, whose practitioners tend to target large specimens.

Birkeland, C. 1997. Implications for resource management. *In: Life and Death of Coral Reefs* (pp. 411–435, Figure 18–2). Chapman & Hall.

Birkeland, C., Dayton, P.K. 2005. The importance in fishery management of leaving the big ones. *Trends in Ecology and Evolution* 20(7): 356–358. https://www.cell.com/trends/ecology-evolution/fulltext/S0169-5347(05)00077-7?cc=yGarcia

118. **Preventing us from hunting/fishing/whaling is causing our youth to leave the area or turn to alcohol and drugs – is that what you want?**

- Making wildlife a scapegoat for social ills is not the answer.
- Killing wild animals is a part of many traditions. In some cases, prohibiting such kills can therefore be considered a curtailment of ancient "rights" or even the rights enshrined in more recent treaties. This, of course, is a hot-button issue considering how often indigenous rights have been trampled upon and the many treaties broken. The issue needs to be considered on a case-by-case basis. Is the hunt really a "make-or-break" cultural foundation? Are the social ills of the local human population perhaps based on multiple factors that require much broader political solutions? Is the hunt a nutritional necessity? Are other foods now available that could compensate for the loss of hunted wildlife? Is the meat of the wildlife in question healthy or potentially highly contaminated (bioaccumulation in top predators such as certain cetaceans)? Why is the hunted species a conservation issue? In all likelihood, that animal population has sunk to some critical level that requires protection. If that is the case, then further hunting can trigger the extirpation of the animal, a worst-case

outcome for the species and certainly not in the ultimate interests of the hunters (see also entry Nr. 113).

Samson, C., Pretty, J. 2006. Environmental and health benefits of hunting lifestyles and diets for the Innu of Labrador. *Food Policy* 31(6): 528–553. https://doi.org/10.1016/j.foodpol.2006.02.001

119. **This species hasn't been sighted here for 50 years and doesn't really belong here anymore, so why shouldn't I shoot it?**

- You mean you don't want the Golden Age of your forefathers, including its natural bounty, to return?
- Shouldn't re-establishing an intact environment and its biodiversity be a reason to rejoice rather than grabbing for your gun?
- Hunting is variously regulated around the globe, but one constant is that certain species are allowed to be hunted in certain numbers during certain well-defined seasons. Other species are on a prohibited list. The training of hunters also ranges from "pay your fee and you're good to go" to rigorous instruction akin to acquiring a vehicle driving licence. Accordingly, hunters may be very poorly informed or even near-professionals. Nonetheless, in both cases, trigger-happiness and the "I'm a hunter, I know better than the pencil pushers" or "wait 'til I show my buddies this trophy" attitude can lead to wrong decisions. One such flawed decision is to shoot an animal that is very rare or that once populated the area and has begun to return in small numbers. Depending on the country, common examples of such "comebacks" include certain deer species, beavers, wild boar, eagles, vultures, lynx, wolves, and bears – animals that were extirpated by hunters in the first place. The current, early, tenuous "successes" are often the result of two different restoration models, namely conservation and rewilding. The former focuses more on endangered species, the latter on promoting natural processes to boost overall biodiversity. All such efforts are typically very time-consuming and expensive, and the relevant authorities are clearly not pleased when hunters shoot the animals. Despite all the public debate surrounding newly reintroduced species, especially if these are predatory, such unauthorized kills are simply illegal. The goals of the relevant experts and authorities – which generally echo the feeling among the general public – are to help restore intact ecosystems for future generations. These endeavors should be supported by hunters, too.

Thulin, C.-G., Röcklinsberg, H. 2020. Ethical considerations for wildlife reintroductions and rewilding. *Front. Vet. Sci., Sec. Animal Behavior and Welfare* 7: 1–10. https://doi.org/10.3389/fvets.2020.00163

120. **This animal is an ancient species, so we believe eating (parts of) it will cure our many ailments and make us live longer.**

- So, eating grass, which is older than any animal, is the healthiest thing you can do?
- We humans have an incredible number of beliefs. They can go by many names, including religions, rituals, dogmas, myths, customs, conspiracy theories, mass hysteria, and delusions. Some are passed down from generation to generation and probably anchored in some purported ancient wisdom, knowledge, or revelation. Others are freshly introduced via an online blog. All warning bells should start ringing when any of the above are based on or advocate killing (and/or eating) certain animals or parts thereof. Especially when the targeted animals are highly endangered. What are we weighing against each other here? Basically, it's some alleged aspect of our well-being versus the extirpation or extinction of a species. In the case of rhino horn, for example, its supposed hedonic value is to treat hangovers. In the case of sea turtle eggs, it's the belief in their aphrodisiac properties. The "ancient" coelacanth (fish) was once rumored (again, now rejected) to provide the raw materials for a "fountain of youth" serum. The scales of scaly anteaters (pangolins) are used to purportedly cure excessive nervousness and hysterical crying in children, women possessed by devils and ogres, malarial fever, and deafness. The species spectrum is enormous, as are the targeted maladies: acne, colds, sore throats, hemorrhoids, conjunctivitis, and even cancer. Beyond the fact that eating many forms of wildlife (bush meat) presents a health hazard in itself (e.g., transmission of zoonotic diseases), few if any such products have ever been scientifically demonstrated to have any effect and, even if so, can be replaced by proven modern medications.

Chinese Medicine and the Pangolin. 1938. *Nature* 141: 72. https://doi.org/10.1038/141072b0

Dang Vu, H.N., Nielsen, M.R. 2018. Understanding utilitarian and hedonic values determining the demand for rhino horn in Vietnam. *Human Dimensions of Wildlife* 23(5): 417–432. https://doi.org/10.1080/10871209.2018.1449038

Senko, J. et al. 2009. To eat or not to eat an endangered species: Views of local residents and physicians on the safety of sea turtle consumption in Northwestern Mexico. *EcoHealth* 6: 584–595. https://doi.org/10.1007/s10393-010-0280-7

121. **This (endangered) species is a predator of other endangered species, so we have to cull it.**

- Sure, as long as we can keep those hunters happy.
- What about culling all the other predators, too?

- Ecosystem or species management can actually involve killing animals. This typically involves culling invasive species or those involved in species introductions gone awry. Examples include rats that have invaded a small island and are killing all the crabs or birds (nests). On the East Coast of the US, abundant raccoons (not threatened) are known to plunder (depredate) the eggs and kill hatchlings from the nests of sea turtles (threatened). Arguments have therefore been made to cull (by trapping) these predators/scavengers. In this case, other, non-lethal, strategies include using various types of screens or cages to protect the turtle nests. Situations in which one endangered species poses a threat to another endangered species are rare and call for detailed study. The "unwanted" predators may, for example, be increasing not because they are consuming the few remaining endangered prey, but because another, over-abundant prey species has become available to them (or because they can also scavenge on the abundant wastes from nearby human settlements). In such cases, one strategy might be to focus on that other prey species (and better manage the household garbage collection system).

Barton, B.T., Roth, J.D. 2010. Raccoon removal on sea turtle nesting beaches. *The Journal of Wildlife Management* 71(4): 1234–1237. https://doi.org/10.2193/2006-014

Gibson, L. 2010. The role of lethal control in managing the effects of apparent competition on endangered prey species. *Wildlife Society Bulletin* 34(4): 1220–1224. https://doi.org/10.2193/0091-7648(2006)34[1220:TROLCI]2.0.CO;2

122. **Nature is so out of balance that we have to hunt to regulate everything – and I'm shooting only one per season, that's sustainable.**
   - Define "balance."
   - And what about regulating the species responsible for unbalancing nature the most – *Homo sapiens*?
   - This is a very common justification expressed by hunters. And it's true that nature often no longer exhibits its original (historical or pre-exploitation) "balance." This is because certain species may have become locally extirpated, introduced species may have spread, or the entire habitat has become compromised by human settlements, agriculture, forestry, and other commercial enterprises. Importantly, however, the concepts of "balance" or "stability" in ecosystems are vigorously debated by experts, with many viewpoints put forward. The issue is currently also being driven by changes due to global driving forces such as climate change. More locally, perceived "unbalances" often involve the human removal of top predators, the selective large-scale removal of dominant herbivores (think buffalo in

the USA), or the hunting-related extinction of other animals (think passenger pigeon in the USA). The motivations in most cases were self-centered, ranging from fear, to the removal of competitors (e.g., for grazing space), or to the removal of perceived pests. Of course, pure sport hunting pleasure must not be underestimated. Is the solution to our having killed off major components of the ecosystem and artificially restructuring food webs simply to kill off other components to achieve some sort of balance? Perhaps only as a last option.

The term "sustainable" doesn't really fit into this discussion: the regulation of hunting (and fishing) is so hotly debated precisely because it is mired by an unending succession of failures. And, crucially, you will never be the only one who is removing just one animal. Animal welfare is also an increasingly important consideration here. Modern ecosystem management has many other tools at its disposal for creating some semblance of balance. These include re-introducing predators or protecting the remaining individuals, establishing national parks, or applying fertility control to prevent the over-population of (non-endangered) competing species. One argument for letting nature take care of the problem is that the populations of overabundant animals tend to regulate themselves once their size exceeds the carrying capacity of the ecosystem.

Bled, F. et al. 2022. Balancing carnivore conservation and sustainable hunting of a key prey species: A case study on the Florida panther and white-tailed deer. *Journal of Applied Ecology* 59(8): 2010–2022. https://doi.org/10.1111/1365-2664.14201

Donohue, I. et al. 2016. Navigating the complexity of ecological stability. *Ecological Letters* 19(9): 1172–1185. https://doi.org/10.1111/ele.12648

123. **You know there were fewer of these animals in historical times, so their numbers are artificially high now and need to be reduced.**

- Well, would you apply that line of argumentation to the human population, too?
- The classical case for that argument is deer in the USA. Extensive recreational hunting, however, may be the poorest option to "reduce," "cull," or "control" populations of wild animals (to mention a few euphemistic terms for those people who don't want to say "kill" outright; see entry Nr. 204). If killing is the only option, then the solution isn't necessarily to involve the general public, that is, those whose sole qualification is to be able to plunk down the fee for a hunting license. Other proposed alternatives are to issue special

permits to kill urban deer, sharpshooting programs by experts to control deer in state parks, or various fertility-control measures. Others include restoring native predators, which affect both the density and behavior of the deer as well as reduce the damage that deer do to the vegetation. Addressing the problem of overabundant deer raises complex biological, ecological, socio-political, public, and legal issues. Better understanding the human dimensions has helped many agencies define appropriate public education programs. This problem will continue to confront wildlife professionals for many decades to come.

Robbi, N.J. 2007. The modern domestic deer hunter: Managing wildlife or wreaking havoc on biodiversity. *37 State Bar of Texas Environmental Law Journal* 150(2006–2007). https://heinonline.org/HOL/LandingPage?handle=hein.journals/txenvlw37&div=37&id=&page=

Warren, R.J. 2011. Deer overabundance in the USA: Recent advances in population control. *Animal Production Science* 51(4): 259–266. https://doi.org/10.1071/AN10214

124. **Fish don't feel pain, so we cut off their fins/descale/gut them while they are still fresh and alive.**
   - Yes, we humans are very good at torturing.
   - This very long-held belief probably arose because most fish do not emit sounds and cannot change their facial expression when captured. The convenient interpretation that fish feel no pain may also reflect self-protection, considering the hundreds of millions of fish that we kill every year, whether hooked, gaffed, crushed in nets, or kept alive under appalling conditions at certain fish markets. The issue has been the topic of numerous scientific publications over the last half-century, with many concluding that fish may not feel pain based on the lack of certain parts of the brain that we humans have. More recent publications that take a more holistic view, however, point to fish experiencing suffering. The fact is that even the lowliest of invertebrates have nerve systems and react to the tiniest of stimuli. We don't doubt that any other vertebrate feels pain, why should fish be an exception? They have well-developed brains and a fully developed vertebrate nervous system. Their violent, uncontrolled thrashing in their death throes should leave little doubt about the stress and suffering they perceive. As in any case when we need to kill an animal, it is simply a sign of disrespect not to do this as quickly and "painlessly" as possible. Even if you believe the scientific evidence is inconclusive, then the precautionary principle would still admonish

proceeding in a manner that causes the least potentially harmful, painful, or protracted outcome.

Braithwaite, V., Huntingford, F. 2004. Fish and welfare: Do fish have the capacity for pain perception and suffering? *Animal Welfare* 13(S1): S87–S92. https://doi.org/10.1017/S096272860001441X

Vettese, T. et al. 2020. The great fish pain debate. *Issues in Science and Technology* 3(4): 49–53. https://www.jstor.org/stable/26949166

8
Pets

125. **I'm saving this animal from its degraded/polluted environment and putting it in my terrarium/aquarium where the conditions are optimal.**

- Permanent captivity never improved anyone's well-being.
- Define optimal.
- Every animal in a cage is not fulfilling its role in its natural ecosystem.
- Two considerations should be priorities in conservation management, in zoos and aquariums, and in the pet industry: animal welfare on the one side, and ecosystem-level aspects on the other. Virtually every form of artificial enclosure, whether it be a cage, a terrarium, or your house (and garden), will be unable to recreate an animal's full environmental spectrum or enable its full range of behaviors. This holds doubly true when an animal is held alone. From another perspective, the exact needs of many animals – ranging from the chemical composition of the water in aquariums to the full behavioral spectrum of the captives – are only partially known. Another key aspect is where the animals came from. In many cases – whether it be for the pet trade or for zoos – animals are captured in the wild and then sold by dealers to customers or to "refresh" captive-breeding stocks. Moreover, many exotic pet species tend to be endangered, or they have become invasive species in their new homes due to escapes or intentional releases ("hey, it simply got too big for my terrarium/cage/living room/bathtub"). Finally, the motivation to hold a pet often reflects a fashion trend or the whims, desires, and needs (sometimes legitimate) of the owner rather than the welfare of the animal. This recognition has more recently begun to alter the pet industry considerably. One example is the ban on displaying certain animals in the shop windows of pet stores. Another is the ever stricter controls and the removal of certain species from the list of animals legally offered for sale.

DOI: 10.1201/9781003757474-8

Sinclair, J.S. et al. 2021. The international vertebrate pet trade network and insights from US imports of exotic pets. *BioScience* 71(9): 977–990. https://doi.org/10.1093/biosci/biab056

126. **I love these wild animals/sea turtles/gorillas . . . why can't I pick them up/pet them/hug them/take a selfie with them to show my affection if I can get close enough?**
    - Your role in life can't be to cuddle wildlife.
    - Would you like aliens to pick you up and squeeze you because they think you're cute?
    - It's an unfortunate phenomenon that modern humans seem to have lost touch with and respect for untamed nature. One reason may be that few of us have ever had the opportunity to see or experience a truly pristine or even near-natural environment. This has led to a disconnect between most people and wildlife. The desire to get up close is perhaps abetted by some wildlife tours that push the permitted limits or by nature films where the camera person and/or professional wildlife managers seem to have moved in very close indeed (even if they whisper into the camera to show their deep respect and understanding of the animals in question, shortly before pulling out the syringe to take a blood sample . . .). This has all been exacerbated by the selfie craze, in which getting up close is the whole point and the only technical option. People who love animals may unwittingly fuel their suffering. Have your picture taken with a tiger? That animal will no doubt be held immobile by a short chain and may well be declawed or even drugged. Sit on an elephant? Pet a wolf? Hug a river dolphin? Smooch with an anteater? Get dressed up in a costume and pose next to a bear? Put it all on Instagram to impress your friends? Simply graceless. The issues are many, ranging from where and how the animals were captured and how they are being held, to habituation, to unexpected (human) injury, to the transmission of diseases (in both directions!). How close is too close? A ranger once told me, regarding sea lions, "if they have to look at you, you're too close."

Bousé, D. 2003. False intimacy: Close-ups and viewer involvement in wildlife films1. *Visual Studies* 18(2): 123–132. https://doi.org/10.1080/14725860310001631994

Geist, V. 2011. Wildlife habituation: Advances in understanding and management application. *Human-Wildlife Interactions* 5(1): 9–12. https://www.jstor.org/stable/24868857

The Hidden Cost of Wildlife Tourism. Special Report, 2019, National Geographic 06.2019.

127. **Yes, it's in a tiny cage, but if I release it, it wouldn't survive a minute out there.**

- A lifetime in a jail cell without having committed a crime . . .
- This situation raises several difficult issues. First, almost any cage will be too small compared with the animal's normal range of activity in the wild. Second, many caged animals tend to show abnormal behavior. Third, many a pet that needs to be caged might involve the "exotic pet trade" or represent a species that poses a threat to its holder (sharp teeth or poisonous). Fourth, those caged specimens may well represent the tiny fraction of wild-caught animals that survived capture and transportation to dealers. Fifth, such animals – and their "family" structure in the wild – may have suffered considerably from the capture (e.g., capturing young animals, sometimes necessitating killing a parent first). Finally, no pet holder should ever consider releasing an animal. While it's true that a pet in captivity is unlikely to die at the hand of a natural predator, this doesn't necessarily outweigh the many negative consequences of a lifetime in captivity. Releasing an animal, however, is almost never an acceptable option. First, that "freedom" is often a traumatic or lethal experience for the animal. Second and more important: official releases by experts involve highly complex and lengthy procedures, are applicable to only a certain subset of animals, and require a correct and largely intact natural habitat. Such programs typically involve a team of experts and volunteers along with long-term supervision (veterinary examinations, pre-release training, post-release monitoring, potentially with tagging). Even then, success is not guaranteed. The take-home message: there will no doubt be some alternative between continuing to keep an animal in a tiny cage and releasing it. That might be a justified and fruitful reason to check the internet for your options.

Lockwood, J.L. 2019. When pets become pests: The role of the exotic pet trade in producing invasive vertebrate animals. *Frontiers in Ecology and the Environment* 17(6): 323–330. https://doi.org/10.1002/fee.2059

128. **This animal is so attached to me, both of us would die of broken hearts if I had to give it away or release it.**

- People should save their undying personal bonds for their mothers.
- The price of a good pet is always a broken heart at the end.
- This aspect depends, of course, in part on the animal's cognitive ability. What may be worthy of discussion in the case of a dog, for example, is no doubt an anthropomorphic misinterpretation when applied to invertebrates, fish, amphibians, reptiles, rodents, and many other

(mammal) pets. In most cases, it would probably be more accurate to refer to a partnership of necessity or convenience, perhaps under the motto, this time taken literally, "don't bite the hand that feeds you." Nonetheless, deep human–pet bonds are an undeniable reality, and many people consider their pets to be family members. Importantly, however, the grief upon premature separation is no doubt deeper and more multifaceted on the human side (see also entries Nr. 127 and 130 regarding releasing).

Milligan, T. 2017. Love and animals. The Oxford Handbook of Philosophy of Love. https://scholar.google.at/scholar?hl=de&as_sdt=0%2C5&q=love+an+animals&btnG=

Tipper, B. 2011. Chapter 8: Pets and personal life. In: May, V. (ed.), *Sociology of Personal Life* (pp. 85–97). Palgrave.

129. **My dog needs exercise – it would be cruel to have it on a leash while walking through this nature reserve, or are you asking me to leave it in the car?! (Figure 24).**

Figure 24 Every dog on a beach will leave behind some bodily wastes, find the garbage, chew on prospective food items, and initiate potentially unfriendly encounters with other beach visitors, whether human or otherwise.

- OK, as long as it doesn't pee, shit, lose any hair, empty visitor trash cans, or interact with any other animal there.

- So, you're putting your dog's welfare above that of nature reserve guidelines?
- Leaving your dog in the car is equally as criminal as taking it into a protected area.
- Most protected areas have strict regulations regarding pets. These have a solid basis in experience and science. First of all, leaving a dog for any length of time in a locked car is, of course, a major no-no. Taking one on a trip to a protected ecosystem, therefore, reflects poor planning, both if it must stay in the car and if the idea was to take it along for the hike. Dogs are typically not permitted inside such areas, whether on a leash or not (with or without a muzzle). Unfortunately, free-roaming dogs are present and pose a problem in many protected areas. The sad truth is also that many dog owners do not have full control of their own pets (and, sadly, therefore no full control over their own lives), especially when these are unleashed and allowed to run free. Otherwise, you wouldn't constantly hear owners ever more loudly shouting their dog's name and/or incessantly whistling. Most unleashed dogs will hark back to innate behaviors and take up the scent of and chase wildlife, whether it be a squirrel or a deer. This is entirely unacceptable. Others will find and pick through trash. Even if leashed, dogs will leave behind any number of excretions that send messages to wildlife and can alter their behaviors. Importantly, the transmission of diseases and parasites between pets and wildlife is a very serious issue. And, finally, if your own health is your top consideration, then you should know that most people killed by cows in Europe are those walking their dogs: such attacks are usually prompted by mother cows pre-emptively protecting their calves from a perceived dog threat.

Aguirre, A. et al. 1995. Wildlife diseases in national park ecosystems. *Wildlife Society Bulletin (1973–2006)* 23(3): 415–419. http://www.jstor.org/stable/3782948

Zamora-Nasca, L.B., Lambertucci, S.A. 2022. Domestic dog-wildlife interactions and support for pet regulations in protected areas. *Biological Conservation* 273: 109705. https://doi.org/10.1016/j.biocon.2022.109705

130. **The animal simply became too large for my aquarium/terrarium/bathtub, and the zoo wouldn't take it, so I set it free.**

- Hope it doesn't eat someone's poodle.
- Park visitors or nearby homeowners will be thrilled to encounter the 2-m-long python you released last year.
- There are many reasons why people become unable or unwilling to continue to properly care for their pet(s). They may involve one of the

many harsh setbacks that we experience over the course of our lives, or simply involve old age (if we have survived those other setbacks). One knee-jerk reaction when a pet has become too large to handle is to simply release it into the surrounding nature. Beyond the admonition that any release be left to experts (see entry Nr. 127), relieving a pet owner of his/her/their responsibility cannot be a justification for any type of release or translocation. The criteria that must be met for official releases include benefiting the respective animal and yielding a measurable conservation benefit at the level of a population, species, or ecosystem. None of these considerations are typically met upon releasing an outsized fish, snake, lizard, turtle, or alligator, to name just a few well-known cases. The issue is compounded by the fact that many of the respective animals are alien or invasive species in their place of captivity. Moreover, based on their size alone, they may have become top predators – even if some may find it mildly amusing to read that a python has eaten an irate poodle. Introduced predators that start killing off the native fauna are a classic example of triggering larger-scale "culling" operations. The state of Florida, for example, has seen fit to hire "python-removal agents." More palatable solutions are to educate prospective owners to make wise pet choices in the first place, to provide information and resources to ensure high standards of animal keeping, and to provide hapless owners with alternatives to releasing. The latter includes giving to a friend, bringing to a humane society or vet's office, donating to a school or zoo, or, as a last resort, having the animal put to sleep by professionals.

Lucca, C.M. 2013. Keeping lions, tigers, and bears (oh my) in check: The state of exotic pet regulation in the wake of the Zanesville, Ohio Massacre. *Villanova Environmental Law Journal* 24: 125. https://heinonline.org/HOL/LandingPage?handle=hein.journals/vilenvlj24&div=7&id=&page=

Reaser, J.K., Meyers, N.M. 2007. Habitattitude™: Getting a backbone about the pet release pathway. *Managing Vertebrate Invasive Species* 40. https://digitalcommons.unl.edu/nwrcinvasive/40

131. **I love animals – I have a cat at home – so you're talking to the wrong person.**
   - Hitler also loved his dog.
   - Being a pet owner isn't a free ticket to contribute to activities that harm wildlife.
   - That's an excuse I heard from a hotel owner who destroyed the sea turtle nesting beach in front of his premises (but apparently had a cat he loved back home). This attitude reflects the increasing disconnect

between people and nature. Many believe they understand the needs of their pets and recognize what their pets do for them but are blissfully unaware of the needs of wildlife or the benefits that wildlife and intact nature may bring to them, either directly or indirectly, or to society in general. In the nesting beach example, the hotel directly benefited from the beach at its doorstep: that's why people booked it (not because of some perceived superior food offerings, exceptional music, or an extra-large bed mattress). The nesting activity itself was also a tourist draw in the area. However, putting out another two rows of uninterrupted sunbeds along with tables, umbrellas, and walkways was clearly a more directly recognizable profit-maker. Ecosystem function (what are those animals doing to keep the ecosystem intact?) and ecosystem services (what are those animals doing that benefits you?) are two concepts that might be introduced if your discussion partner happens to have an open ear. And, of course, talking money is the language that everyone understands, most of all the business people that nature conservationists are most often confronted with "on the other side of the negotiating table." Unfortunately, biologists and conservationists tend to have the least grasp of money and the financial world, probably due to their pitiful pay scales. Start learning to "talk money" now!

Baveye, P.C. et al. 2013. Monetary valuation of ecosystem services: It matters to get the timeline right. *Ecological Economics* 95: 231–235. https://doi.org/10.1016/j.ecolecon.2013.09.009

132. **I heard that an alligator/snake once killed a pet poodle. We can't have that!**
    - Do you prefer TV programs about alligators and other wildlife, or those about poodles?
    - "The poodle bites, the poodle chews it." (Frank Zappa)
    - I see that you and I actually read the same grocery store magazines.
    - Ok, it's no doubt true that an alligator or snake somewhere and at some time actually killed a pet. Is that in any way surprising? Alligators and larger snakes are very clearly top predators, and many can't really be faulted with trying to eat anything that passes their snouts or doesn't stop yapping at them. The time to get really worried is when such an animal has its eyes set on you. Whether the potential prey is you or your chihuahua, the first take-home message is never to release your reptile just because it has grown too large for the terrarium or bathtub. The second message is that you and your hopefully leashed pet should stay clear of such large reptile predators. In some cases that

means not taking your dog along for your wetland excursion. Other situations may actually involve what is referred to as a "nuisance" or "problem" alligator. A quick call from your cell phone will have the appropriate authorities heading your way to solve the problem in a jiffy. Their usual strategy: quickly capture the beast and relocate it unharmed elsewhere (or possibly kill it if it is an alien or invasive species) – all at no cost to you.

Hines, T.C., Woodward, A.R. 1980. Nuisance alligator control in Florida. *Wildlife Society Bulletin* 8(3): 234–241. https://www.jstor.org/stable/3781695

133. **My one big parrot – I love it and it loves me.**
   - "Love means never having to say you're sorry." (Love Story).
   - Almost all parrots are social animals, whereby many form monogamous pairs within their flock. Keeping a single parrot is therefore a recognized no-no today. It can lead to major behavioral anomalies ranging from aggression (or apathy) to feather plucking. If your bird really "loves" you, that is, sees you as its mate, then that's another seriously misguided and poorly reversible disorder known as misimprinting. That means a lifetime of frustration for your poor bird (and need not necessarily be pleasant for you). One guiding principle would be to keep such parrots as a male–female pair from the beginning (after ensuring that they are, in fact, compatible). And, of course, only if you can provide them with the larger aviary (and the appropriate food and environmental enrichment objects) that fulfil modern pet-keeping guidelines. Note also that a parrot means a lifelong commitment: it may well outlive you.

Luescher, A.U. (ed.). 2006. *Manual of Parrot Behavior*. Blackwell Publishing. 332 pp. https://doi.org/10.1002/9780470344651

van Hoek, C.S., Ten Cate, C. 1998. Abnormal behavior in caged birds kept as pets. *Journal of Applied Animal Welfare Science* 1(1): 51–64. https://doi.org/10.1207/s15327604jaws0101_5

134. **My parrot's cage is quite big, so I'm taking perfect care of it.**
   - How would you like to spend your whole life in a one-room "tiny house"?
   - Birds fly, and parrots often fly considerable distances. No cage can ever be large enough or simulate the wide-open skies. Yet cage size is only one criterion. Even if you add environmental enrichment structures and objects, no cage can simulate the natural environment (e.g., sunlight exposure) and promote normal foraging and mating

behavior, let alone regular contact with flock members. This can lead to stereotyped behaviors, that is, those that are repeated over and over again without any apparent adaptive function, such as running the beak up and down cage bars or tossing the head back and forth. That points to extreme boredom and suffering. Many owners will fail to fully recognize these symptoms or misinterpret them. When the parrot then starts to pluck its own feathers, whether out of boredom, sexual frustration, or other stress factors, it's often too late. Finally, only a few parrot keepers will be able to recreate the natural diet, which varies considerably between and within species and can be variously based largely on seeds, nuts, fruits, or vegetables. Over the many decades that your parrot can live, meeting these needs ends up being quite expensive.

Kim, L.C. et al. 2009. Preferences of Orange-winged Amazon parrots (*Amazona amazonica*) for cage enrichment devices. *Applied Animal Behaviour Science* 120(3–4): 216–223. https://doi.org/10.1016/j.applanim.2009.06.006

Peng, S., Broom, D.M. 2021. The sustainability of keeping birds as pets: Should any be kept? *Animals* 11(2): 582. https://doi.org/10.3390/ani11020582

9
Zoos and Public Aquariums

135. **Don't worry, if the population sinks too low, we'll just catch the last animals and breed them in zoos.**
    - So, you want to remove the last individuals from their environment, too?
    - Zoos are not really designed to re-establish historical population levels.
    - This is a common misconception on several levels. Firstly, once a population sinks to a certain very low level, its genetic makeup tends to be highly uniform, leading to the so-called "bottleneck effect," whereby subsequent interbreeding produces a population that is not broadly set up to withstand the full range of natural diseases. Secondly, not all species can be successfully captured and then captive-bred. Some suffer lethal stress during capture (e.g., the failed attempt by an international expert team to capture one or two of the last vaquita porpoises in the Gulf of California (Mexico)). Thirdly, while some species can be kept in captivity, not all will successfully breed. This has led to procedures such as artificial insemination, which does not necessarily improve the welfare of the animals involved. In extreme cases (e.g., certain rhinos), eggs or sperm must be taken from aged or even deceased animals in an attempt at in vitro fertilization ("test-tube rhinos," carried to term by surrogates of the species' closest relatives). All such attempts require a long-term commitment and are extremely costly. Wouldn't it simply be much more reasonable to save them (and their habitat) before such a critical situation arises?

Keulartz, J. 2015. Captivity for conservation? Zoos at a crossroads. *Journal of Agricultural and Environmental Ethics* 28: 335–351. https://doi.org/10.1007/s10806-015-9537-z

Rojas-Bracho, L. et al. 2019. A field effort to capture critically endangered vaquitas *Phocoena sinus* for protection from entanglement in illegal

DOI: 10.1201/9781003757474-9

gillnets. *Endangered Species Research* 38: 11–27. https://doi.org/10.3354/esr00931

136. **Sure, that lion is forced to pace back and forth in its cage, but at least it doesn't have to fight to defend its territory.**
    - But territoriality is only one of a dozen big cat behaviors that zoos should be enabling.
    - Zoos are increasingly striving to enable as many of the behaviors of a zoo animal as possible – and to avoid stereotyped behavior (e.g., incessantly pacing back and forth). This begins with so-called "behavioral enrichment," in which animals are provided with various structures as environmental stimuli. These can be elevated perches, climbing opportunities, or various toys they can play with and tear apart. It can also involve placing food in more difficult-to-reach areas of their enclosures or requiring them to make a bigger effort when feeding (for cheetahs, for example, meat suspended on a pully system that is rapidly drawn across their enclosures, forcing them to run). More recently, decisions have been made to enable large cats to successfully mate and produce cubs, even if those cats are abundant in zoos and there is little chance that the cubs can be accommodated elsewhere in the future. This means that the cubs may ultimately be killed but indicates that zoos recognize the need to ensure that the animals can display the largest number of behaviors possible to ensure their well-being in captivity. And, of course, that zoo strategy is not entirely altruistic or animal-welfare focused. After all, cubs are a proven public attraction, bringing in crowds (and profits). . .

    https://www.sciencedirect.com/topics/veterinary-science-and-veterinary-medicine/behavioral-enrichment

137. **If I can see it at the zoo, I don't really care about its status in the wild.**
    - So, you want to make zoos into museums?
    - So, it's all about you, is it?
    - The rarely publicly stated truth is that zoos often rely on wild populations to resupply their animals. Some animals are directly wild-caught, others are the products of captive-breeding programs within and exchanges among zoos. Ensuring the genetic diversity of captive populations, however, can require fresh animals from the wild. This is one reason why it's in the zoos' own interest to make sure that viable populations exist in nature. Moreover, many zoos state that part of their mission is to ultimately release captive-bred individuals back into the wild (even if this may be a rare occasion involving very few species due to the considerable long-term costs in time, effort, and money).

Such projects are especially important when wild populations are declining or have been extirpated. The technical term here is reintroduction. The primary aim can therefore be to promote the recovery of a single endangered species. On a broader scale, however, ecosystem restoration is often an equally important goal. It's no use reintroducing an animal into an environment that cannot fulfil its needs. This is the case especially when a keystone species is being protected or reintroduced, one that influences other species or plays a role in the function of the overall ecosystem. Regardless of the perspective, it is certainly in everyone's interest that wild populations and intact ecosystems exist.

Polak, T., Saltz, D. 2011. Reintroduction as an ecosystem restoration technique. *Conservation Biology* 25(3): 424–425. https://doi.org/10.1111/j.1523-1739.2011.01669.x

Taylor, G. et al. 2017. Is reintroduction biology an effective applied science? *Trends in Ecology and Evolution* 32(11): 873–880. https://www.cell.com/trends/ecology-evolution/fulltext/S0169-5347(17)30191-X

138. **These lions lie around all day anyway, so whether they do it in a cage or in the wild is really all the same.**
   - If you were a prisoner in solitary confinement, you'd probably be lying around a lot too.
   - So, if they are lying around, like in nature, then the zoo must be doing a fantastic job, right?
   - Lions, as all animals, show a wide range of innate behaviors and live in complex social units (prides). No zoo can fully recreate such a large animal's natural environment, give rise to and maintain a fully adequate pride (sometimes exceeding two dozen animals and including related lionesses, males, and cubs), or enable the full range of behaviors such as roaming and hunting. Very poor keeping conditions, for example, in cages of any sort, lead to abnormal or stereotyped behavior such as restless pacing (cage stereotypy), a sign of boredom and stress. Zoos are making major strides in providing more space and stimulation (enrichment). For large cats, this can include nutritional enrichments such as suspending food from trees or attaching it to fast-moving platforms that require them to exhibit certain hunting behaviors. Other efforts to improve the cat's environment can include hanging robust "lion balls," burlap bags or other toys, providing wooden perches, leaving dung or scent trails of blood or urine, or wrapping rough ropes around tree trunks to promote self-grooming.

Goswami, S. et al. 2021. Effects of a combined enrichment intervention on the behavioural and physiological welfare of captive Asiatic lions (*Panthera leo persica*). *Applied Animal Behaviour Science* 236: 105222. https://www.sciencedirect.com/science/article/pii/S0168159121000095

139. **OK, so you're taking good care of the animals here in the zoo, but what are you doing for the populations in the wild?**

- Everything humanly possible considering our limited opportunities and funds.
- Zoos around the world have faced various criticisms, mostly involving animal welfare considerations. Importantly, all zoos recognize that they depend, at least for certain animals and at certain intervals, on animals captured in the wild. This has led zoos to expand their missions and improve their image by increasingly focusing on wildlife in the respective countries of origin. This can involve:
  - financially supporting wildlife conservation projects in the wild;
  - sending experts to help train local wildlife veterinarians and rangers;
  - initiating research projects in the field;
  - promoting public information campaigns in the respective countries;
  - initiating studies to examine the behaviors of animals in the wild and thus help to enable as many such behaviors as possible in zoo enclosures;
  - inviting experts from the respective countries to the zoo for further training and education; and
  - working at captive breeding programs that aim to reintroduce animals into their original environment.

Accordingly, public zoos and aquaria are increasingly committed to conservation, nature restoration, sustainability, and education. This involves collaborations with organizations such as the World Wide Fund for Nature (WWF) and associations such as the World Association of Zoos and Aquariums (WAZA), the European Association of Zoos and Aquaria (EAZA), and the European Zoo and Aquarium Association (EUAC). This helps define common goals, divide up the tasks, and unite the institutions in meeting them. Such efforts are one step toward helping animals that are endangered or already extinct in the wild in order to preserve the species.

Correia, J. et al. 2024. A review of two decades of in situ conservation powered by public aquaria. *Journal of Zoological and Botanical Gardens* 5(1): 90–118. https://doi.org/10.3390/jzbg5010007

Miranda, R. et al. 2023. The role of zoos and aquariums in a changing world. *Annual Review of Animal Biosciences* 11(1): 287–306. https://doi.org/10.1146/annurev-animal-050622-104306

140. **That's a very dangerous animal – it's better here in a zoo cage than running around in the wild.**

- Actually, it's just as dangerous in the zoo, too.
- The word "dangerous" is problematic when applied to animals. Every animal has defense mechanisms to enable it to face up to members of its own species, to prevent being preyed upon by a predator, or to prey on others itself. Some use toxins (venomous snakes, stinging insects, jellyfish), others have sharp weapons (teeth, claws), and still others can use their own weight or muscle as a weapon. Not to mention the chemical defenses of many plants, some of which are even potentially lethal to human beings. This means that virtually any animal much larger than a chihuahua (and some much smaller) can pose a real threat to you and me. Conversely, the word "tame" should never be used in conjunction with a zoo animal.

  And speaking of chihuahuas, you and your loved ones are much more likely to be injured by a dog than by any animal in the jungle. That neighborhood dog may have been purposefully raised to be ill tempered, for example, as a guard dog. Or it may be basically benign but simply happens to have a toothache, be plagued by parasites, to have just lost its rank in the hierarchy, be sleep-deprived, or be skittish around children that pull its tail, blow in its face, or try to poach its food. Or you have provoked its ire by being a mailman, by trying to kiss its cute nose, or by unwittingly triggering its predatory instincts by running away.

  Importantly, complex protocols are in place in zoos (and subject to regular review, especially after "mishaps") to prevent injury or death of zoo employees (and to protect zoo visitors!). Note also that dangerous animals tend to be public attractions (for armchair thrill-seekers?) that boost zoo profits. Clearly, the answer cannot be to attempt to eliminate all such species in the natural environment and keep a few specimens for our viewing pleasure in zoos. Of course, many of those "dangerous" animals play an important role in their ecosystem, making them so-called keystone species, whereby some are also highly "attractive" so-called flagship species. The latter can attract funding, and their protection in the wild may help protect other species in that habitat or even the ecosystem itself.

Murphy, H.W. 2011. Dangerous animal crisis management. *Fowler's Zoo and Wild Animal Medicine Current Therapy* 7(7): 78. https://books

.google.at/books?hl=de&lr=&id=18A5OdzUBqEC&oi=fnd&pg=PA78&dq=dangerous+animal+zoo&ots=Gu4f8oz7zW&sig=WXMxkD2uv2MQP9LHGlt3SeuK4ng#v=onepage&q=dangerous%20animal%20zoo&f=false

141. **What's wrong with knocking on the aquarium glass to attract their attention?**

- How would you like someone ringing your doorbell every 2 minutes?
- Knocking on the glass front of exhibits is a considerable disturbance for fish, most of which are very sensitive to loud acoustic stimuli. Noise disturbance can alter fish behavior and physiology. The rates measured at one public aquarium were two knocks per 100 visitors, amounting to up to several hundred per day. Another study reported that two to five knocks per day killed 90% of one fish species. With thousands of visitors a week, that adds up quickly. Management strategies include information signs ("Knocking kills fish"; "Only the daft tap on glass"; "Please don't knock on the glass") or an additional protective glass pane between visitors and the aquarium itself. If you really want to "interact" with aquarium animals, see whether the aquarium might have a "touch tank." At this point, it might be worth mentioning that aquarium staff have repeatedly observed children placing their hands on the front panel of a public aquarium and spreading their fingers out to "enlarge" the image or to zoom in on a particular animal. This is a sad statement about the triumph of the digital over the natural world and highlights the effort that needs to be made to bring nature closer to the younger generation.

Knudson, H.M. (2019). *Empathy for Invertebrates: Adults' Empathic Behaviors at Aquarium Touch Tanks* (Master's thesis, Univ. of Washington). http://hdl.handle.net/1773/44638

Kopczak, C. et al. 2013. Families talking about ecology at touch tanks. *Environmental Education Research* 21(1): 129–144. https://doi.org/10.1080/13504622.2013.860429

Kratochvil, H., Schwammer, H. 1998. Reducing acoustic disturbances by aquarium visitors. *Zoo Biology* 16(4): 349–353. https://doi.org/10.1002/(SICI)1098-2361(1997)16:4<349::AID-ZOO7>3.0.CO;2-A

Rowe, S., Kisiel, J. 2012. Family engagement at aquarium touch tanks – Exploring interactions and the potential for learning. In: *Understanding Interactions at Science Centers and Museums: Approaching Sociocultural Perspectives* (pp. 63–77). Rotterdam: SensePublishers. https://brill.com/display/book/edcoll/9789460917257/BP000006.xml

142. **I paid my entrance fee and that animal looks skinny. Why can't I feed it candied popcorn?**
- Paying a fee doesn't entitle you to improper behavior.
- How would you like it if someone fed your child candy all day?
- Zoo visitors can have negative, neutral, or positive effects on zoo animal behavior and welfare. On the negative side, the desire to feed zoo animals is a vexing problem in zoos worldwide. All the more unpleasant because the intentions are good, but the results are potentially lethal. In some zoos and for a particular species, pre-packaged food tailored to that species may be sold in small amounts to visitors. Otherwise, feeding is strictly prohibited. On the one hand, this is because the animals quickly learn to associate visitors with food, leading to unnatural and unwanted behavior. More importantly, this ban is because most species have very specific diets and should only receive a certain amount of that per day. Popcorn, bread, and other human snack remains are certainly not a part of any animal's natural diet. The most common strategy to combat uncontrolled feeding is to put up "do not feed" signs, although these may be ineffective if not accompanied by explanatory information. Other strategies include establishing a barrier or a larger distance between visitors and the animals.

Bitgood, S. et al. 1988. Controlling public feeding of zoo animals. *Visitor Behavior* 2(4): 6. https://scholar.google.at/scholar?hl=de&as_sdt=0%2C5&q=feeding+zoo+animals&oq=feeding+zoo

Sherwen, S.L., Hemsworth, P.H. 2019. The visitor effect on zoo animals: Implications and opportunities for zoo animal welfare. *Animals* 9(6): 366. https://doi.org/10.3390/ani9060366

143. **I love how that elephant stands there and sways its head from side to side – it's dancing for me.**
- It's trying to tell you something, but I'm afraid it's not melodious.
- You're totally wrong, it's not a "dancin' fool." (Frank Zappa)
- Unfortunately, what looks like a dance is actually a sign of mental distress related to captivity and is one type of abnormal stereotyped behavior, similar to pacing back and forth in cages by captive large cats. Stereotypy is defined as any movement performed repeatedly, relatively invariant in form, and with no function or goal. In elephants, it can include weaving (swaying the body from one side to the other and swinging the trunk), head bobbing (repeated up-and-down or forward-and-back rocking of the head, and pacing (moving a foreleg from one side to the other or hind leg forward and back

and swinging the trunk while standing still. The prevalence of swaying apparently increases in individuals kept largely indoors and those with lengthier periods of being chained. Being deprived of maternal relatives and isolation from other elephants promotes the behavior. Other contributing factors include no access to outdoor enclosures or to natural feeding, little opportunity for walking or exercise, or daily routines that are timed for monotonous work year-round. Preventing such negative keeping conditions and enabling a wide range of behaviors underlines the incredible costs and efforts associated with keeping elephants in zoos, not to mention the ever-present dangers posed to zoo staff members. Environmental enrichment strategies to combat stereotypy can include foraging opportunities, a pool, interactive toys, sufficient social contact with other elephants, and even music and other sounds.

Rees, P. 2020. *Elephants under Human Care: The Behaviour, Ecology, and Welfare of Elephants in Captivity*. Academic Press. 380 pp. https://books.google.at/books?hl=de&lr=&id=g6f_DwAAQBAJ&oi=fnd&pg=PP1&ots=2WkNY5UzcA&sig=ReD1301EljBremPagJj2uxevafI&redir_esc=y#v=onepage&q&f=false

Rees, P. 2024. Stereotypic behaviour in elephants. *In: Animal Behaviour and Welfare Cases*. https://doi.org/10.1079/abwcases.2024.0002

Vanitha, V. et al. 2016. Prevalence of stereotypies and its possible causes among captive Asian elephants (*Elephas maximus*) in Tamil Nadu, India. *Applied Animal Behaviour Science* 174: 137–146. https://doi.org/10.1016/j.applanim.2015.10.006

10
Natural History Museums

144. **Museums should be closed – all the animals and plants there are dead.**
   - So, where else will you take your kids on the weekend?
   - OK, but stuffed animals smell better than living ones.
   - Museums are a most valuable repository of all things natural (and even cosmic – think meteorites). Not even all zoos in the world combined could hold the inventory of species of one well-endowed museum, even if those specimens are all preserved in one way or another rather than being alive. In fact, there is often little overlap at all between what museum goers can see and what zoos exhibit. Visit both! Many of the species they hold in their collections are inconspicuous and never make it into the public eye; many are perhaps better known only to experts in the field. This includes many insects and other invertebrates. Importantly, every species that has been given a scientific name is based on a collected specimen: preserving that individual or those individuals is a crucial factor in verifying new specimens, recognizing slightly differing species, and in making decisions if name changes are suggested (e.g., based on an even earlier description or if the larger taxonomic affiliation was originally entirely incorrectly interpreted). And, of course, some of the specimens, such as the dinosaurs, are long extinct and have been painstakingly reconstructed by experts found only in museums. Those experts are also often the ones going into the field to collect and save valuable samples that would otherwise be lost (think road construction and dinosaur bones). Overall, the collections are ideally suited to promote a better understanding of the diversity of life – past and present – on our planet.

Mujtaba, T. et al. 2018. Learning and engagement through natural history museums. *Studies in Science Education* 54(1): 41–67. https://doi.org/10.1080/03057267.2018.1442820

DOI: 10.1201/9781003757474-10

145. **Museums collect dozens or even hundreds of specimens of the same species. That's total overkill.**

- The number of individuals held in museum collections is dwarfed by the numbers killed heedlessly by human activities over the centuries.
- Those specimens from decades or even a century or more ago have gone a long way to describing the variability in a particular species or even documenting species that are no longer present in the region. Each specimen is, of course, unique in its genetic makeup but typically also in other features such as age, size, shape, sex, or coloration. This is important in determining differences both within and between species. Its place of capture can help determine past and present ranges of distribution. Having a number of specimens also allows a range of detailed investigations, some of which require dissecting or otherwise damaging a specimen. Moreover, specimens were often collected at different times in the past, helping to quantify crucial developments such as changes in pollutant levels over time (such as the mercury content in bird feathers). Finally, having many individuals enables curators to send specimens for scientific research to other museums or research institutions for study. Interestingly, many researchers from universities and elsewhere ultimately donate their collections to a museum for permanent safekeeping and future research. And, as experience shows, future research typically involves entirely novel methods and technologies (genetics, electron microscopes) that reveal completely new levels of detail and information.

  https://lymanmuseum.wordpress.com/2011/10/28/why-so-many-specimens/

146. **Museums are an anachronism.**

- If so, then why are they so hopelessly overcrowded every weekend?
- Museums have a greater public outreach than almost any other institution I can think of.
- In fact, museums are constantly evolving to better address the interests and challenges of modern society. A key challenge in managing collections is to optimize the value they give to society, both now and in the future. Throughout, they seek to maintain a balance between research, teaching, and public information. Today, museums are powerful research institutions on par with universities, and their staff publish their results in a wide range of scientific journals. Some museum professionals are also teachers and provide training in their laboratories. In many cases, their labs and technical facilities are the only avenues to making fossil finds visible and rendering organisms in their full beauty, whether prepared or as models. Importantly, a major

mission is to educate the public, which they do through their exhibitions, public lectures, guided field trips, and the like. Recently, they have become key players in promoting citizen science (i.e., projects partially or entirely involving amateurs or members of the interested public in collecting and reporting data). The presentations to the general public are constantly updated to incorporate the newest technical innovations. This helps guarantee a good measure of entertainment (think interactive displays, moving dinosaurs, etc.). Overall, this makes museums key motors in making society aware of Earth's past, the present challenges, and potential future scenarios. All these considerations point to a bright future for museums.

Dorfman, E. (ed.). 2017. *The Future of Natural History Museums*. New York: Routledge. 248 pp.

Waller, R., Henderson, J. 2024. The problem of compromise in conservation and exhibit decision making. *Collections* 20(2): 298–312. https://doi.org/10.1177/15501906241232454

11
Pollution

147. **Oil is a natural product – the spill can't be all that bad.**

- Well, heroin and cocaine are also natural products, not to mention snake venom.
- One valid mantra here is "Sola dosis fecit venenum" (the dose determines whether something is a poison or not). That phrase was coined five centuries ago, showing that we have long understood the basics of toxicity. A classic example is salt. Almost no meal anywhere in the world would be complete without a dash of it. But several tablespoons at once is a lethal dose. Of course, some substances are highly toxic in the very smallest of doses. Some such substances are artificial and not normally present in the natural environment (and organisms may react accordingly upon contact). Others may be highly concentrated and overwhelm physiological processes (not only a pollution problem but also relevant in drug overdoses).

The amount of crude oil spilled into the world's oceans is estimated to lie somewhere between five million and ten *million tons* per year. That's a big dose. Crude oil can contain up to several thousand different compounds, some of which are toxic in even small amounts. More importantly, oil spills involve enormous amounts that can coat and decimate entire coastlines. And, importantly, some of the "oil" being transported and applied around us on a daily basis is not crude oil extracted from the earth but variously refined oil products, most of which have a higher toxicity. Major, regular tanker accidents such as those of the Torrey Canyon, Amoco Cadiz, or Exxon Valdez have gone down in infamy. The same holds true for major oil platform blow-outs (remember the Deepwater Horizon oil spill in the Gulf of Ameri . . . er, I mean Gulf of Mexico, in 2010). Oil kills by:

- being toxic;
- smothering organisms;

DOI: 10.1201/9781003757474-11

- causing seabirds to freeze because the insulation of their feathers is lost;
- causing seabirds to drown because the feathers are no longer "waterproof";
- being ingested and clogging the respiratory or digestive tracts; and
- by hindering reproduction (e.g., contaminating bird eggs in nests or damaging sea turtle nesting beaches).

Who is responsible? The oil company? The shipbuilders? The inadequate "flag-of-convenience" regulations that some oil tanker owners adhere to? The ship captain? Or is it also you and me, who demand petroleum products to power our vehicles and heat our homes?

Bhattacharjee, S., Dutta, T. 2022. Chapter 1: An overview of oil pollution and oil-spilling incidents. In: Das, P. et al. (eds.), *Advances in Oil-Water Separation* (pp. 3–15). https://doi.org/10.1016/B978-0-323-89978-9.00014-8

Shaw, B. et al. 1987. The global environment: A proposal to eliminate marine oil pollution. *Nat. Resources J.* 27(1987): 157. https://scholar.google.at/scholar?start=20&q=oil+pollution+problems&hl=de&as_sdt=0,5

148. Hey, it's only a paper cigarette butt, and I smoked outside, so get off my back! (Figure 25).

Figure 25 Cigarette butts, by far the most common beach litter all around the world, are tiny toxic bombs (and made largely of plastic, not paper!). And where you find one, you're likely to find many in what might be called the "law of marine litter aggregation." This says a lot about most smokers.

- If it's so small and only paper, why can't you take it home and dispose of it properly?
- Oh yeah, drop one in your aquarium at home and see what happens to the fish.
- People who leave their butts behind should have their behinds butted. (Mike Stahovich).
- Cigarette butts are the most frequent litter in the world and always top the lists of items collected during beach clean-ups. Six trillion cigarettes (that's a 6 with 12 zeros!) are smoked every year, and most are improperly disposed of. Contrary to popular belief, each cigarette filter is composed of plastic, not paper, and can take years to decompose. Moreover, after a smoke, each one contains many hundreds of toxic substances, some carcinogenic. If you don't believe me, drop one or two of these little toxic bombs into your aquarium at home and see what happens to the fish. This is yet another reason why smoking is banned not only in many buildings, but also increasingly outdoors, including on many beaches. And it has also triggered high-level discussions on banning cigarette filters, which would be designed to reduce the palatability and appeal of cigarettes and to correct consumer misconceptions that filters significantly reduce health risks.

Nowotny, T.E. et al. 2009. Cigarette butts and the case for an environmental policy on hazardous cigarette waste. *International Journal of Environmental Research and Public Health* 6: 1691–1705. https://doi.org/10.3390/ijerph6051691

149. **Every volcano spews out a thousand times more pollutants than this factory.**
   - Maybe, but we have many factories and no volcanoes here.
   - You need to distinguish between natural catastrophes and human-made ones.
   - That's a common thought, but it is based on a fundamental misconception: although volcanic eruptions are often said to pollute the environment, they are not typically considered to represent pollution in the strict sense. For example, the United Nations defines marine pollution as: "The direct or indirect introduction *by humans* of substances or energy into the marine environment (including estuaries), resulting in harm to living resources, hazards to human health, hindrances to marine activities including fishing, impairment of the quality of seawater for various uses and reduction of amenities." This clearly excludes substances emitted by natural events or disasters. Obviously, we are not responsible for volcanic activity and cannot do anything to control it. We can, however, reduce the waste

that we ourselves produce, and factories are a prime example. It's also important to differentiate what exactly is being spewed out by the two sources. As opposed to volcanic emissions, anthropogenic inputs often involve synthetic substances that are unknown in nature and for which no biological decomposition processes are in place (so-called "forever chemicals"). Beyond directly toxic substances, manufacturing industries are also responsible for more than one-third of global CO_2 emissions. Overall, the emissions of most factories can be reduced to acceptable levels. Advanced technologies and, after the fact, even biological approaches (phytoremediation: using plants to reduce concentrations or toxic effects of contaminants in the environment) can play a role here. Of course, it's also a matter of money and priorities. Are clean air and water low on your list of priorities?

http://data.un.org/Glossary.aspx?q=marine+pollution

Verschoor, A.H., Reijnders, L. 2000. Toxics reduction in ten large companies, why and how. *Journal of Cleaner Production* 8(1): 69–78. https://doi.org/10.1016/S0959-6526(99)00316-9

150. **This water looks pretty clean to me, I think I can drink it?**

- Looks can kill.
- I'm sure you can – once . . .
- Maybe let someone else in the group you don't like try it first.
- This is a question that typically comes up in every lengthier outdoor excursion. And the answer will typically be "no, it's not worth the risk." Why? Because pollutants know no borders and most can be detected even in the farthest corners of the globe (albeit usually at low concentrations there). From a more practical perspective, none of the most hazardous pollutants in aquatic systems are visible. This is the case for heavy metals, radioactivity, and most other industrial chemicals, including pesticides and pharmaceuticals. Importantly, contamination of the water with (invisible) bacteria, viruses, and parasites continues to be a major scourge responsible for countless deaths worldwide every year. Such pollution can be suspected only when sewage or other effluents visibly enter and cloud the water. Do you know whether animal herds might be grazing (and defecating) at a higher elevation than you are now? Or maybe visitors regularly camp at or above the tree line? All their excretions will ultimately run downhill, in your direction. Even outdoor recreational facilities may fail to provide microbially uncontaminated drinking water. Is there a positive take on bacterial contamination in this respect? Well, the reactions in your digestive system (e.g., gastroenteritis) usually

take effect much faster than the symptoms of other pollutants. That might give you some advanced warning if you can afford to wait to see whether others fall ill first. Unfortunately, a large portion of the human population is supplied with drinking water that is iffy at best. If you're a camper or hiker, it might pay to take along a pack of water purification tablets. A large range of portable backpacking water filters and treatment systems is also available. Of course, a low-tech option is to boil the water first (one minute at least).

Phiri, B.J. et al. 2021. Microbial contamination in drinking water at public outdoor recreation facilities in New Zealand. *Journal of Applied Microbiology* 130(1): 302–312. https://doi.org/10.1111/jam.14772

Senthil Rathi, B. et al. 2021. Critical review on hazardous pollutants in water environment: Occurrence, monitoring, fate, removal technologies and risk assessment. *Science of the Total Environment* 797: 149134. https://doi.org/10.1016/j.scitotenv.2021.149134

151. **The atmosphere and oceans are endless. Everything is diluted to insignificant amounts.**

- How much do you want your sewage to be diluted before you drink it?
- Even our poorly developed noses can detect compounds such as those in diesel fuel at very tiny concentrations in the parts per million (ppm) range (just open the door to the engine room of any larger boat and take a whiff). If you can smell it, how much more will a dog or a fish be irritated?
- Even if pollutants are diluted in water or the atmosphere to parts per million or parts per billion (ppb), we can usually measure them. And if we can measure them, then most organisms can detect and be impacted by them. Many animals can detect even the tiniest amounts of substances. A few molecules can suffice for the decisive clue, for example, to find a partner or to orient toward a faraway food source (how do you think flies appear so quickly?). The millions of tons of pollutants we spew into the environment can overwhelm and cripple such fine-tuned detection systems.

On the one hand, pollutant inputs into a particular location in the sea are not necessarily distributed widely: many remain more strongly concentrated in the water or sediments around the waste disposal site itself. Although the oceans could be said to all be interconnected, they are not one contiguous basin with perfect mixing and dilution. Even if they were, however, a back-of-the-napkin calculation that might be called the sugar-cube paradigm is a real eye-opener: All the world's oceans together contain 10^{21} liters of water. A cube of sugar contains 10^{22} molecules of glucose. Accordingly, if that one cube could be equally dissolved in the overall volume of the marine environment (and the glucose would not decompose), every liter of water would contain five molecules of glucose (Egon Degens). You don't have to be a mathematician to imagine what that calculation looks like for the millions of tons of effluents we introduce into the sea every year.

On the other hand, even if the pollutants become more widely distributed, some can then again be accumulated in certain regions by ocean currents. Case in point: the accumulation of plastic litter in the center of several large ocean basins by circular surface gyres (buzzword: "the Giant Pacific Garbage Patch").

From yet another perspective, many organisms can take up and accumulate the tiniest amounts of toxic substances (bioaccumulation) to amounts much higher than in the surrounding water. These levels can be harmful and then pose an even greater threat to predatory fish that eat lots of such contaminated prey (biomagnification). In the case of heavy metal pollution, for example, we as seafood consumers can be harmed via that process. In fact, heavy metals such as mercury are one of the few forms of marine pollution that have actually killed people via that route (Minamata "disease").

Hachiya, N. 2006. The history and the present of Minamata disease – Entering the second half a century. *Japan Medical Association Journal* 49(3): 112–118.

12
Energy

152. **We'll just solve the windmill (and energy) problem by putting them all out at sea: no eyesore, no one is bothered by the noise, no valuable land is lost.**

- And if we install them at regular intervals, maybe the whales can use them as a slalom course.
- Wind energy is, in fact, one of the promising solutions to help reduce our dependence on petroleum products as an energy source. Windmills on land have led to pushback for esthetic and environmental reasons (massive cement bases, noise, the blades kill birds and bats, and indigenous rights). Offshore windmills were therefore deemed a logical alternative and are part of the marine renewable-energy industry. Experience, however, has shown that windmills at sea also have considerable drawbacks. The first is that their erection, servicing, and later dismantling are considerably more expensive than for their onshore counterparts. Environmentally, the noise generated during their construction and operation poses a threat to marine organisms, in particular whales and dolphins. The same holds true for the vibrations they emit. The altered electromagnetic fields associated with their myriads of submarine cables are also thought to affect fish and their distribution. Importantly, windmill construction inevitably requires heavily modifying the underlying seafloor (with certain later potential benefits subsumed under the term "reef effect"; see entry Nr. 60). Their aggregation in windfarms also poses physical barriers to sea mammal migrations. The Environmental Concerns Subcommittee of the International Whaling Commission, for example, has recognized offshore windmills as a threat in two categories: habitat modification and anthropogenic noise. Lastly, these facilities can also pose a hazard to marine traffic. The US government recently canceled approval for two large windfarm projects off the coast of New England. A political

DOI: 10.1201/9781003757474-12

or environmental decision? For the role of policy-makers in nature conservation, see entry Nr. 160.

Riefolo, L. et al. 2016. Offshore wind turbines: An overview of the effects on the marine environment. Paper presented at The 6th International Ocean and Polar Engineering Conference, Rhodes, Greece, June 2016. https://onepetro.org/ISOPEIOPEC/proceedings-abstract/ISOPE16/All-ISOPE16/17779

Simmonds, M.P., Brown, V.C. 2010. Is there a conflict between cetacean conservation and marine renewable-energy developments? *Wildlife Research* 37: 688–694.

Wang, L. et al. 2024. Ecological impacts of the expansion of offshore wind farms on trophic level species of marine food chain. *Journal of Environmental Sciences* 139: 226–244. https://doi.org/10.1016/j.jes.2023.05.002

153. **If you wouldn't oppose damming this river for hydroelectric power, we wouldn't need that damn nuclear power plant.**

- So, you're arguing for sacrificing a river so that you can power your electric toothbrush?
- How about trying to reduce our electricity consumption instead of producing ever more?
- In fact, every form of energy production, even those touted as being "renewable" or "green" (solar, wind, biomass, geothermal), has its drawbacks. In this case, the comparison is between a major change to an ecosystem (the river) and a millennia-long toxic waste disposal problem (nuclear plant). Rivers are among the world's most distinctive features, determining the positions of major cities, the livelihoods of many residents, and serving as key transport routes. Their function both for human well-being and as key ecosystems, however, is being increasingly compromised by pollution, shoreline regulation, and water diversions. A key parameter describing their role and health is "connectivity," meaning a river's link with itself in the direction of flow and laterally with adjoining habitats such as riverine forests, wetlands, and floodplains. Damming fragments rivers and compromises this connectivity. Ecologically, this alters their entire dynamics, including water flow as well as the composition and movement of species, nutrients, and sediments. Beyond altering such ecosystem functions, damming impairs the benefits (ecosystem services) they supply to us, such as fisheries, buffering of flood events, recharging groundwater, healthy deltas, along with recreational and cultural values. Accordingly, rather than damming the few remaining naturally flowing rivers and constructing

new dams on the others, efforts are increasingly being made to reverse the fragmentation of flowing waters and restore their connectivity by removing artificial barriers. This calls for carefully weighing and promoting other forms of sustainable energy production before turning to alternatives such as nuclear power with its proven catastrophic outcomes in case of major accidents and the "forever" commitment to guard its toxic wastes.

Thieme, M. et al. 2024. Measures to safeguard and restore river connectivity. *Environmental Reviews* 32: 366–386. https://doi.org/10.1139/er-2023-0019

154. **A nuclear power plant here will reduce our dependency on oil and gas and also help combat climate change.**
   - Yes, and the next 100 generations of your family will have guaranteed jobs guarding the nuclear waste.
   - The most erroneous stories are those we think we know best – and therefore never scrutinize or question. (Stephen Jay Gould)
   - You're right, it'll be a "glowing" future.
   - The promise of nuclear energy has been hotly debated for decades. The fact that efforts are being made to declare it a "sustainable" or "green" energy source (in the EU) has misled many into believing that it can solve our energy problems in an environmentally friendly manner. The main argument is that it would reduce our dependence on petroleum products and thus help reduce the CO_2 output that arises when fossil fuels are burned. Any significant effect in this regard has largely been debunked and is dwarfed by the negative and often unsolved or unsolvable problems. These include:
     - the fuel, uranium, is a limited resource and therefore also unsustainable;
     - uranium mining is a health hazard;
     - the cost of building nuclear power plants is exorbitant;
     - planned new nuclear power plants would not be operational for at least a decade;
     - past accidents have proven to be major long-term disasters, and future accidents *will* happen;
     - nuclear power plants could be targeted in armed conflicts (and such a plant has recently ended up on the front lines of a war);
     - transporting nuclear wastes between facilities (sometimes internationally) poses a risk;
     - producing nuclear fuels increases the risk of nuclear proliferation (the spread of nuclear weapons);

- the issue of long-term nuclear waste disposal has never been satisfactorily resolved and is probably unsolvable. From that perspective, any so-called closed "nuclear fuel cycle" is a myth; and
- nuclear fission? Not in your lifetime. Don't believe the hype.

All truly more sustainable alternative strategies should be implemented first.

Müller, W.C., Thurner, P.W. (eds.). 2017. *The Politics of Nuclear Energy in Western Europe*. Oxford University Press.

Nuclear energy is not a "green" investment. https://www.bmk.gv.at/en/topics/climate-environment/nuclear-coordination/nuclear-not-green.html

Weish, P. 2006. Tuning in: Energy at the turning point – From oil to sun. *In: Nuclear Power, Climate Policy and Sustainability: An Assessment by the Austrian Nuclear Advisory Board*. Federal Ministry of Agriculture and Forestry, Environment and Water Management, 21–26. https://scholar.google.at/scholar?hl=de&as_sdt=0%2C5&q=Peter+Weish+nuclear+power&btnG=

155. **You flew here in an airplane to participate in this conservation project/conference: is that ecologically justifiable?**
   - If it gave me an opportunity to talk with you person-to-person, it was worth it.
   - That's a legitimate concern. There is no simple answer. In the case of scientific symposia and conferences, efforts are now being made almost everywhere to reduce their number or to replace some of them (or parts of them) with virtual communication technologies. Sometimes, centralizing a symposium location at the regional scale or, alternatively, decentralizing large international conferences over multiple sites can be options to reduce distances flown or to replace air flight with other means of transportation. In some cases, annual meetings are being replaced by biennial ones (e.g., International Whaling Commission). In other cases, for example, when high-level negotiations are required or where expert groups need to tackle a complex technical issue together, an in-person meeting may be necessary.

     The situation is different for most conservation projects. In contrast to symposia, which tend to last only a few days, volunteers and other participants in conservation projects tend to put in weeks to months of strenuous fieldwork. In such cases, local capacities or manpower may be missing, and only minimal or no action would be taken without outside input. Here, the benefits for the endangered species or the threatened habitat (and for the respective country) probably outweigh any sustainable travel considerations.

Caset, F. et al. 2018. Meet & fly: Sustainable transport academics and the elephant in the room. *Journal of Transport Geography* 70: 64–67. https://doi.org/10.1016/j.jtrangeo.2018.05.020

156. **I see you researchers have vehicles, drones, computers, lamps, energy-intensive high-tech equipment, refrigerators and stoves, cell phones, and bottled water, so you're not setting a very good, environmentally compatible example.**
    - Energy and tools used to help save our planet have clear priority over other uses, don't you think?
    - Yes, research and fieldwork today often require mobility and energy-consuming devices. This is the only way to make such fieldwork possible and effective. Importantly, such modern equipment and the amenities help reduce the number of people we would otherwise need, greatly reducing our footprint. Keeping the members of our team healthy and fit is paramount. Forgoing or withholding certain amenities could ultimately incur much greater costs and put a greater burden on local resources. Finally, we are doing all this not only for ourselves or for science but also for our/your country. We hope that the environmental costs are reduced by having chosen the optimal location for our work. Those costs are also balanced by the new knowledge gained, by training and transferring skills to local people, and by boosting overall public awareness. Together, this will help improve future conservation efforts. We would, of course, be grateful if you could help us in any way to be more environmentally friendly.

https://www.rgs.org/research/higher-education-resources/fieldwork-principle-2-sustainable-fieldwork

157. **So you're against nuclear energy, against damming rivers for hydroelectricity, against windmills on land and at sea – do you want to take us back to the caveman era?**
    - I certainly wouldn't want you to part ways with your beloved electric toothbrush.
    - If the caveman era was so abominable, why are man caves so popular?
    - That's a tough question. Global energy demand is on the rise. We clearly need ever more energy to power our daily lives and to keep businesses and industry humming. The answer can only be two-fold: seek the most environmentally friendly sources of energy and make all attempts to reduce energy consumption. The first includes further expanding the use of solar energy, above all by installing panels on already sealed surfaces such as the roofs of homes and businesses. Geothermal energy would be another ecofriendly option wherever

possible. Equally important is reducing our electricity consumption by using ever better appliances and electronic devices. Are the horrendous amounts of energy soaked up by booming data centers and the cryptocurrency market really optimally advancing human well-being (data centers, for example, currently consume more than 20% of Ireland's electricity and have caused water shortages elsewhere)? Of course, a large-scale, global switch to electric vehicles will put another major dent in efforts to reduce electricity consumption. Other promising counter-strategies include environmentally friendly architecture, insulating existing homes, and a more economical use of energy (that starts with simply turning off unneeded lights). Regular feedback to consumers by energy providers can also play a role.

Dickson, M.H., Fanelli, M. 2013. *Geothermal Energy: Utilization and Technology*. Routledge. First published 2003 by United Nations Educational, Scientific and Cultural Organization (UNESCO), Paris, France. ISBN 1-84407-184-7. https://scholar.google.at/scholar?hl=de&as_sdt=0%2C5&q=Dickenson+Fanelli+Geothermal&btnG=

Fischer, C. 2008. Feedback on household electricity consumption: A tool for saving energy? *Energy Efficiency* 1: 79–104. https://doi.org/10.1007/s12053-008-9009-7

13
Climate Change

158. **Global warming can't be a problem considering how much I froze last winter.**
    - Everybody talks about the weather, but nobody does anything about it.
    - So, what's your game plan: rebuild or relocate?
    - Experts say we need to distinguish between the weather and the climate. Perhaps it's a bit like "two sides of the same coin," whereby the weather is basically the short-term condition and climate is its long-term average. This means the local weather is determined by factors such as regional meteorological events, geography, and time of year, but referring to the climate involves filtering out all those ups and downs to get the broader, more long-term picture. Accordingly, a cold spell experienced in the winter should not be interpreted as a counterargument to global warming. The virtually unanimous consensus of researchers around the world is that the climate is changing. The reports by the responsible international body – the Intergovernmental Panel on Climate Change (IPCC) – indisputably document a rapid and ongoing change in the world's climate and clearly attribute that to human causes. Some factions of the general public still wish to debate whether this change is natural or human-driven, whereby it is informative to check where in society the "natural causes" proponents are anchored. Others try to skirt the issue by pitting the emissions of one source against those of another, for example, the trucking industry against the cement industry, or the automobile sector against air traffic. That's not the answer: every sector must do its best to reduce emissions. Still others put their hopes in what has been termed "the myth of carbon capture and storage (CCS)," whereby carbon (carbon dioxide) is removed from the atmosphere and pumped into huge "reservoirs" deep underground or below the seafloor (don't hold your breath). Regardless of how you might feel about this issue, climate

DOI: 10.1201/9781003757474-13

change is harming the human population (not to mention ecosystems worldwide). You and your family should get ready for some of those changes. Take your pick: disappearing coastlines, unusual heatwaves, extended drought, extreme forest fires and wildfire smoke, drying lakes, increased flooding, stronger storm activity, shifting animal populations (think mosquitoes)... one or the other is bound to affect you or an upcoming generation. Accordingly, climate change action is one of the greatest health (improvement) opportunities.

Hamilton, L.C. et al. 2015. Tracking public beliefs about anthropogenic climate change. *PLoS ONE* 10(9): e0138208. https://doi.org/10.1371/journal.pone.0138208

Intergovernmental Panel on Climate Change (IPCC). https://www.ipcc.ch/reports/

Schneider, S.H. 2011. *Encyclopedia of Climate and Weather* (Vol. 1). Oxford University Press. https://books.google.at/books?hl=de&lr=&id=av7q4N8Ib6sC&oi=fnd&pg=PP2&dq=weather+versus+climate&ots=fQvUe7go64&sig=zipY8W8aJEdaiA5wJudoG8BC8HE#v=onepage&q=weather%20versus%20climate&f=false

159. **The world's climate has changed many times in the past, and the world is still standing, so why should I worry!**

- What, me worry? (Alfred E. Neuman)
- Yes, it's true that our planet has gone through many cycles of major climate change (and also experienced major extinction "events"). Natural changes, however, are measured in scales of many millennia and longer, whereas human-induced changes take place over years, decades, and centuries at most. The slow rates of natural change enable animal and plant species to adapt, either genetically or by migrating along with the changing boundaries of their original habitat. Genetic adaptation is unlikely in anthropogenically driven changes, at least for the animals that most folks know, and migration may be thwarted by the many barriers that civilization has put up (not to mention natural barriers such as the northern limits of certain seas like the Mediterranean). Yes, the world will remain "standing" regardless of any severe climate change, but it may not be a world you'll recognize or find comfy. For you personally, I guess it partly depends on what you expect from nature.

Stern, D.I., Kaufmann, R.K. 2014. Anthropogenic and natural causes of climate change. *Climatic Change* 122: 257–269. https://doi.org/10.1007/s10584-013-1007-x

160. **Climate change is just a fake issue so that scientists can get more money.**
    - Better to be safe than sorry.
    - Scientists (especially those in biological fields) are no doubt among the most poorly paid experts on the planet, regardless of any funds that might be made available for their research.
    - Beyond local or indigenous people, scientists are often the first to recognize an environmental problem. And scientists are clearly the first to quantify such problems. Due to the so-called science–policy gap, however, it can take some time for decision-makers and the general public to become more fully aware of an issue or to take action. Several potential explanations for this problem can be forwarded:
        - there is a time gap (sometimes years) between obtaining research results and their publication in peer-reviewed journals;
        - no politician or policy-maker has ever read a scientific paper in a research journal;
        - politicians and policy-makers would like to ignore advice on environmental degradation and impending species extinctions because the solutions always require money;
        - there are competing or vested interests, some with much stronger lobbies;
        - accordingly, politicians and policy-makers seek and cling to counter-arguments, often of iffy provenance, in the hopes of delaying any costly measures;
        - many environmental problems are long-term issues, but the politicians' terms of office are comparatively short ("4 years"), so let the next candidate take care of it; and
        - finally, even if the science is convincing, many folks fail to "believe" in climate change for two reasons. The first is that our own personal actions are decoupled from the climate effects in both time and space: some of us may only begin to suffer many years from now, and those who might suffer sooner may be far away. The second is because the effects are "unintentional" and often caused by invisible substances.

    Once a problem is more broadly recognized, especially when human welfare is imperiled, then the submission rates for relevant research projects tend to increase and be funded. This can give the impression that many researchers are jumping on the bandwagon and attempting to get a "slice of the cake." Of course, it's true that working successfully on an emerging hot topic is one doorway to a rewarding career, but this does not detract from the importance or urgency of a topic. Based on the recognized interdependence of climate, ecosystem health, biodiversity, human well-being, and sustainable development, an increasing range of scientific fields have become involved in climate action. Their assessments are based on facts, and their predictions are being made with an increasing level of confidence. The flow of money in their direction may soon make the future of your family brighter. And don't forget, the truth will come out sooner or later (hopefully sooner). After all, the Brazilian president labeled a recent climate summit "the conference of truth."

Lee, H. et al. 2023. *Synthesis Report of the IPCC Sixth Assessment Report (AR6), Longer Report*. IPCC. https://mural.maynoothuniversity.ie/id/eprint/17733/

161. **If you're so worried about climate change, why did you fly down here for this conservation project?**
   - We wanted to get here as fast as we could to improve your future.
   - Yes, that's a legitimate question. Practically, based on our limited funds, you can be sure that every project and trip is planned meticulously and that no unnecessary flights are approved. We are also making an effort to recruit and involve local people in conservation efforts here rather than flying in personnel from our organization/university/company. That would also be important in the event of any future travel restrictions. At some point, we hope that our work can be taken over entirely by local groups. Philosophically, each one of us here today basically hopes to do more good than damage.

   Nekaris, K.A.I. et al. 2022. From international to local: Promoting local volunteer tourism to guarantee the persistence of wildlife conservation projects in the post-COVID-19 era. *COVID* 2(9): 1287–1302. https://doi.org/10.3390/covid2090095

162. **Climate change? I look forward to having much warmer summers.**
   - Maybe it'll get hot enough to poach eggs in your pool.
   - Considering all the hoopla about immigration, are you willing to pay to make new homes in your neighborhood for the people displaced from low-lying islands?
   - "It's like a heatwave; It's burning in my heart; I can't keep from burning; It's tearing me apart." (Martha and the Vandellas)
   - Warmer summers might be the case for certain people in certain regions of the world. Unfortunately, your warmer summers could well become someone else's lethal heatwaves. Or catastrophic fires. Or

major storm and flooding events. Some areas may become unlivable. Entire habitats (and even island nations) may be lost. Regardless of where you are or your connection with nature, you'll have to fend for yourself. And you'll have to use your own money to protect yourself and your family – don't expect major long-term bailouts from the government. Farmers, for example, may have to abandon their farms, city folk, especially the elderly, will experience higher levels of mortality in their homes and apartments, and may no longer be able to go outside during the day in summer. And it wouldn't be wise to place your hopes on some future technological fix for increasingly severe climate calamities (see entry Nr. 17).

So do your share now: eat more meat, buy exotic fruits and vegetables, sell your bicycle, pump more gas into a bigger car, turn your heating up/your air conditioning to the max (T-shirts in winter, sweaters in summer!), pour that cement and asphalt, approve that nearby mega-data center so that AI can make your life wonderful and your crypto-currency transactions go smoothly, and then fly regularly to ever more far-away destinations to get away from it all. Wait a minute – did I get that right? Or have I been misinformed like Rick (Humphrey Bogart), who wisecracked "I came to Casablanca for the waters". . .?

Guo, Y. 2017. Heat wave and mortality: A multicountry, multicommunity study. *Environmental Health Perspectives* 125(8). https://doi.org/10.1289/EHP1026

Shi, K., Touge. Y. 2022. Characterization of global wildfire burned area spatiotemporal patterns and underlying climatic causes. *Scientific Reports* 12: 644. https://doi.org/10.1038/s41598-021-04726-2

Yadav, N. et al. 2023. Systematic exploration of heat wave impact on mortality and urban heat island: A review from 2000 to 2022. *Urban Climate* 51: 101622. https://doi.org/10.1016/j.uclim.2023.101622

14
Construction Projects and Zoning

163. **Stop nagging – frankly, this area looks a lot more beautiful than it did just a few years ago, thanks to our re-zoning law and the new landscaping.**
- Beauty is in the eye of the beholder.
- You mean a parking lot is beautiful?
- Someday the blind will see, and the deaf will hear . . . the message that we are paving over the very environment that enables our life and well-being on this planet.
- There are two types of sites that the powers that be may seek to "beautify." The first are certain natural habitats that some may view as unspectacular and not particularly worthy of protection. The second involves the many natural habitats that we have turned into fallow or wasteland or that have become abandoned commercial sites. The latter category encompasses land that was probably despoiled by ill-advised zoning in the first place (from nature to commercial?). Sometimes, the reclassification or redesignation of such land may lead to a landscape that many will find more appealing than the earlier status quo. A new plantation, a new golf course, or a new shopping mall are arguably more esthetic or less threatening than a deserted farmstead, a razed shanty town, or an abandoned mining site. That comparison, however, is flawed. It's merely comparing two land zoning strategies – one representing clearly failed or unsustainable former land uses, the other merely promising to cover up those failures. It's basically cosmetics. And equally as shaky as comparing apples and oranges.

Overall, such planning and zoning can contribute to inequities, environmental injustice, and unfavorable risk profiles ("riskscapes"). "Ecological zoning" may be one promising approach here. Yes, some of the "consumer-oriented" new land uses may provide less unappealing and potentially less toxic surroundings, much like a cemetery no

DOI: 10.1201/9781003757474-14

doubt provides a less objectionable soundscape than a demolition derby dirt track, drag-racing strip, or a recreational off-road vehicle and motocross training site. The truly valid comparison, however, would be between the original ecosystem and what we have ultimately made out of it. That is seldom done because virtually no commercial use would ever prevail from an environmental perspective (as project proponents well know and therefore keep mum about – or try to cover up with greenwashing). Moreover, the original condition is off almost everyone's radar screen. This is one aspect of the shifting baseline syndrome, which in simplified form states that each human generation accepts the situation in which it was raised as being normal, failing to recognize (or not knowing) the original, pristine ("pre-exploitation") condition – and therefore failing to take the necessary action at the necessary time.

Kai Li, K. et al. 2024. Can ecological zoning act as an environmental management tool for protecting regional habitat quality: Causal evidence from the national key ecological function zone in China. *Journal of Cleaner Production* 475. https://doi.org/10.1016/j.jclepro.2024.143623

Wilson, S. et al. 2009. How planning and zoning contribute to inequitable development, neighborhood health, and environmental injustice. *Environmental Justice* 1(4). https://doi.org/10.1089/env.2008.0506

164. **This shopping mall will be built on a filled-in swamp, which is a win-win situation for everyone.**

- A win for everyone except the swamp fauna and flora.
- Beware of the politician who wants to "drain the swamp."
- "Drain the swamp" is a timeless phrase repeatedly used by politicians. Although it typically refers broadly to reducing bureaucracy or corruption, it harks back to physically draining swamps to uproot something deemed bothersome (e.g., mosquitoes, alligators). In fact, it is a common practice to use zoning and other measures to encroach on "undesirable," unlucrative, or untapped habitats and landscapes and turn them into more profit-bringing plots. Only today are the functions and roles of many formerly disregarded or unappreciated landscapes being recognized – in the case of swamps, their status as places of beauty in and of themselves, their promise as tourism attractions, and as sites for numerous human recreational activities (e.g., canoeing, fishing, photography). Not to mention their key roles in water purification and flood control. Once seen as wastelands, they have now become national treasures. All warning bells should ring when developers argue for converting remaining natural or semi-natural

sites into commercial landscapes. And if you're more of a money person, then it's very unlikely that any significant benefits or profits will go directly to you. In the case of shopping malls, for example, the typical course of events is initial commercial success followed by failure and abandonment as soon as the next bigger shopping mall opens up nearby. More generally, many "brick-and-mortar" establishments are facing hard times and closures due to new (online) shopping habits. Rarely will the land under those failed facilities be returned to its pre-construction state.

Scharoun, L. 2011. Utopia lost? The significance of the shopping mall in American culture and the effects of its decline on the American public. *Australasian Journal of Popular Culture* 1(2): 227–245. https://doi.org/10.1386/ajpc.1.2.227_1

165. **This business park/industrial complex/major residential project will have some solar panels and greenery on its roofs, making it fully eco-friendly (Figure 26).**

Figure 26 A major effort was made to prevent the further erosion of this beach remnant by planting dune vegetation (despite the sign, even footprints are forbidden in this fenced-off area) – but some witless authority approved the habitat-truncating and defacing construction, and some uninspired architects designed the hideous result.

- "You can fool all of the people some of the time and some of the people all the time, but you cannot fool all the people all the time." (Abraham Lincoln/P.T. Barnum)
- I'll believe you if you give me one of the free T-shirts with your logo and a cute animal on it that you are passing around.
- A new building complex will typically cover a plot of land that once featured naturally grown vegetation and served as a home for the associated fauna. The subsequent landscape beautification between, around, and on top of the buildings can never replace that habitat. Importantly, additional large surfaces will typically be sealed with asphalt for employee, customer, or residential parking and walkways. Finally, the massive amount of cement used in the construction phase presents a climate-related problem of its own (the construction industry is responsible for nearly 25% of all greenhouse gas emissions). The future resource requirements for operation (water, cooling, heating, waste removal, etc.) are rarely factored in or even mentioned in the glossy brochures produced for prospective customers. Public relations efforts to pass massive construction projects off as environmentally friendly can only be a classic case of greenwashing, which has become a central component in global economy and policy strategies. Are the developers passing the project off as environmentally friendly because it's close to public transportation? Are they touting sustainability because they envision five building lots to symbolize the "5 elements of nature" (I'm not kidding, the worst combination of misguided esoterism and greenwashing I've come across)? **Don't believe the hype!** Prick up your ears if contractors try to "sweeten the deal" with promises of a day care facility, bakery, or other amenity to lull and silence local opposition.

He, Q. et al. 2020. To be green or not to be: How environmental regulations shape contractor greenwashing behaviors in construction projects. *Sustainable Cities and Society* 63: 102462. https://doi.org/10.1016/j.scs.2020.102462

Rehan, R., Moncef, N. 2005. Carbon dioxide emissions and climate change: Policy implications for the cement industry. *Environmental Science & Policy* 8(2): 105–114. https://doi.org/10.1016/j.envsci.2004.12.006

166. **This golf course project will create a green oasis and be a healthy recreation mecca.**
    - But golfers wear really weird pants.
    - "They call it golf because all the other four-letter words were taken." (Ray Floyd)

- Golf courses are among the most severe encroachments on a habitat (besides mining) and have some of the worst environmental "report cards" of any landscape. On the one hand, the original habitat and terrain are almost always completely obliterated with heavy machinery to create the topography necessary for the sport (tees, fairways, greens, roughs, obstacles, mounds, sand traps, water hazards). On the other hand, most golf courses today are conceptualized as golf clubs or exclusive resorts with extensive infrastructure, including apartments, restaurants, expansive parking areas, etc., so that large peripheral surface areas are permanently sealed. Other environmental problems include:
  - planting of non-native grass species;
  - excessive water use to maintain the grass surfaces (a crucial issue in arid regions);
  - use of herbicides to prevent the growth of weeds;
  - use of fertilizers to maintain the grass surface and other (planted) vegetation;
  - use of pesticides to keep those pesky insects at bay for the well-heeled clientele; and
  - in alpine countries, appropriation of the limited flat countryside.

In fairness, the ecological value of golf courses regarding biodiversity is being discussed, with a certain degree of "improvement" possible in areas that were previously severely degraded. Nonetheless, golf requires more land per player than any other sport. Finally, golf courses are by definition (fore!) also off-limits for recreational or any other uses by the general public.

Colding, J., Folke, C. 2009. The role of golf courses in biodiversity conservation and ecosystem management. *Ecosystems* 12: 191–206. https://doi.org/10.1007/s10021-008-9217-1

Guzmán, C. et al. 2014. Environmental impacts by golf courses and strategies to minimize them: State of the art. *International Journal of Arts & Sciences* 7(3): 403–417.

15 Money and Jobs

167. **What you are asking for is going to kill my job!**
    - If a conservation project poses a threat to your job, then your job probably poses an unsustainable threat to the environment.
    - And what about my job in nature conservation?
    - Modern environmental programs and conservation projects tend to consider humans as part of the ecosystem and therefore make every effort to incorporate human concerns. Of course, this cannot mean preserving jobs or creating new jobs in enterprises with proven major negative impacts on the environment. Often, however, modes of business can be changed with little effort to become more environmentally friendly (e.g., reducing beach furniture and light pollution on sea turtle nesting beaches), which means no loss of jobs. Importantly, many conservation projects are designed to maintain intact nature, and that in itself has proven to be a major attraction in the framework of the tourism industry and therefore a generator of new, stable, healthy local jobs. Most folks choose their vacation destinations based on natural beauty, not necessarily on gross national product or some exceptional service or amenities that are not to be found anywhere else in the world. Accordingly, nature conservation efforts worldwide create or safeguard millions of jobs in ecotourism and sustainable fishing alone and help considerably reduce atmospheric CO_2 emissions.

Claes, J. et al. 2020. Valuing nature conservation. *In: A Methodology for Quantifying the Benefits of Protecting the Planet's Natural Capital.* McKinsey & Company. https://scholar.google.at/scholar?hl=de&as_sdt=0%2C5&q=Claes+Valuaing&btnG=

Wolf, K.L., Housley, E. 2016. Young adult conservation jobs and worker health. *Journal of Environmental Planning and Management* 60(10): 1853–1870. https://doi.org/10.1080/09640568.2016.1262833

DOI: 10.1201/9781003757474-15

168. **What you environmentalists are asking for is much too expensive.**

- Nothing is more expensive than a missed opportunity. (Horace Jackson Brown, Jr.)
- The four most expensive words are "We don't need nature."
- The best things in life are free – if we keep the environment healthy.
- Compare it to the price of a fighter jet or ballistic missile.
- Sometimes an effort to help save a species or a habitat is inexpensive, sometimes expensive. In both cases, it can be very labor-intensive and time-consuming (that's why "volunteers" often play such an important role – see entry Nr. 212). Even if that 'manpower' is expensive, it's creating jobs and lasting value, and you're treading on mighty shaky ground these days if you're against "jobs." The return on investment can be enormous. One study estimated that doubling nature conservation on land and in national waters by 2030 would create approximately 400,000 to 650,000 jobs in conservation-management fields such as wildlife management and area infrastructure. That study continued by stating that "Through adjacent nature-dependent markets, natural capital could also support local economic growth, generating or safeguarding on the order of $300 billion to $500 billion in GDP and 30 million jobs in eco-tourism and sustainable fishing alone." To remain in those financial dimensions, environmental goods and services provided by natural biological systems have been valued at between $2.9 and $33 *trillion* annually. In contrast, the estimates for the efforts to maintain biodiversity cluster around a comparatively low $20 billion. That sounds like a fantastic return on investment (you don't have to remember the exact sums, just the orders of magnitude involved). Importantly, preserving the environment at an early stage of deterioration costs only a tiny fraction of that for repairing major damage or attempting a restoration at a later date. Considering these sums, one (only slightly) extravagant adage by a well-known scientist regarding seeking funds for research was "Don't ask for millions, ask for billions" – that'll grab their attention and give them an inkling of the gravity of the task and what is at stake.

Claes, J. et al. 2020. Valuing nature conservation. *In: A Methodology for Quantifying the Benefits of Protecting the Planet's Natural Capital* (52 pp.). McKinsey & Company. https://scholar.google.at/scholar?hl=de&as_sdt=0%2C5&q=Claes+Valuaing&btnG=

James, A. et al. 2001. Can we afford to conserve biodiversity? *BioScience* 51(1): 4–52. https://doi.org/10.1641/0006-3568(2001)051[0043:CWATCB]2.0.CO;2

169. **We only need a tiny part of this land/nature reserve for our commercial enterprise. What's the problem with that?**

- That's akin to "slow slicing" or "death by a thousand cuts" (a historical form of torture and execution).
- Give them an inch, and they'll take a mile (give them a centimeter, and they'll take a kilometer . . .).
- This is a typical line of argumentation for choosing commercial interests over intact nature (besides purportedly creating jobs). Such reasoning also often involves a healthy portion of greenwashing. Closer examination, however, will almost always reveal that the land destined to be built over is itself only a remnant of the former natural ecosystem. A classic example in Europe, for example, was the plan to destroy only a few percent of the Danube River floodplain forest for a major hydroelectric dam project in the early 1980s. In fact, the remnant forest and overall floodplain at that time represented only a very tiny fraction of its historical extent. This "only a fraction (of a faction)" argumentation is a typical "death by a thousand cuts" fate in many conservation versus business scenarios. That argument for the Danube could only be brought forward due to the shifting baseline syndrome (see entry Nr. 163), in which the public had little idea of the original extent of what today is considered a highly valuable ecosystem and part of which has become a coveted National Park. The dam project was thwarted, in large part by student protests, in what marked a turning point for environmental awareness in German-speaking Central Europe and was of great significance for the development of democratic processes in Austria.

https://en.wikipedia.org/wiki/Occupation_of_the_Hainburger_Au

Schoder, A., Schmid, M. 2017. Where technology and environmentalism meet. *In: The Remaking of the Austrian Danube for Hydropower. Environmentalism in Central and Southeastern Europe: Historical Perspectives* (pp. 3–20). Lexington Books. https://books.google.at/books?hl=de&lr=&id=xbg9DgAAQBAJ&oi=fnd&pg=PA3&dq=Schoder+Schmid+hydropower&ots=gt6izxq2La&sig=HAHPD-RqnHo_YX1-H0WoxlcDr9w#v=onepage&q=Schoder%20Schmid%20hydropower&f=false

170. **You want to use my hard-earned taxpayer's money to clean up the air/save this species/restore this habitat? I think there are hundreds of better/more important ways to spend the money.**

- Speaking of clean air – when you can't breathe, nothing else matters. (American Lung Association)

- You mean you'd rather spend even more on missiles, drones, and bombs (or on further expanding the nation's highway network)?
- Destroying the country's natural environment might leave you with a lot fewer taxpayers in the future to pay for the things you think are more important.
- It certainly is a tricky task to divide up taxpayer money among the many demands placed by modern society. That might be one reason to be a biologist rather than a politician. Where can the money do the most good? Where does it have the most promise for long-term, sustainable outcomes? No side is ever satisfied (well, maybe sometimes, secretly, the military . . .). We biologists should have no fear of talking about money. Unfortunately, biologists are relatively poor at talking about it because, well, most are on the poorer side and don't have a lot of experience in dealing with that dirty concept! Our task starts with being able to put price tags on nature and the services it supplies, or at least being able to define financial orders of magnitude. More specifically, it involves naming and balancing the costs of a conservation-related effort with the (often hidden and delayed) cost of *not* doing that project. In a larger context, much of the global economy depends on natural capital – the world's stock of natural assets. Human activity is eroding that natural capital. This calls for major, ongoing investment in conservation (and, ha, maybe more money for biologists). The earlier we tackle critical environmental issues, the less we will have to spend to ensure our well-being and survival in the future. And since you mentioned taxes, many creative strategies are available. One approach involves so-called environmental taxes designed to help replace damaging human activities with more sustainable alternatives, for example, to reduce pollution. Another tax-related instrument is to support conservation by providing incentives to promote voluntary, permanent restrictions on private land use through conservation easements.

Parker, D.P., Thurman, W.N. 2018. Tax incentives and the price of conservation. *Journal of the Association of Environmental and Resource Economists* 5(2): 331–369. https://doi.org/10.1086/695615

Ye Xu et al. 2023. Impact of environmental tax on pollution control: A sustainable development perspective. *Economic Analysis and Policy* 79: 89–106. https://doi.org/10.1016/j.eap.2023.06.006

171. **We will make a major financial contribution to your nature protection organization if you don't officially object to the business/factory we want to build here.**

- So, you would rather negotiate than do what's right from the start?
- You do know I have the microphone on, don't you?
- What do you mean you're holding my dog to ransom until this is satisfactorily resolved for you?
- In the real world (by which many mean the business world), when large sums of money are involved, things can get "dirty" pretty fast. Commercial enterprises realize that major projects in the landscape face increasing hurdles these days. One such hurdle is push-back from environmental groups. Commercial enterprises may counter this with slick PR and greenwashing efforts presenting such projects as being ecofriendly ("hey, we will have paper straws") and as providing all sorts of benefits to the area, or emphasizing their corporate social responsibility policies ("if you allow us to build 120 units, we'll see if we can squeeze in a day care room").

One approach that businesses use to dampen the opposition to their upcoming, potentially environmentally threatening projects is to contact selected environmental groups in advance for behind-the-scenes discussions to sound out the potential objections and the fervor behind them. Every type of negotiation anywhere at any time involves some measure of give-and-take. Business directors have a 'duty' to keep costs low and maximize share value, which often means offloading as many of their costs onto other people or the environment as they can. They come to the table with "optimal" demands. Environmentalists should also counter with maximum demands. Unfortunately, they sometimes lack confidence and tend to arrive at negotiations with a compromise already in hand, ultimately leading to a further "compromise on a compromise." Potential enticements by businesses for cooperation by your environmental organization may be tempting. Don't rashly take the bait. Even if the deal can help strengthen your organization, beware of the strings attached (do you have a lawyer?) and avoid potentially suboptimal negotiating outcomes for conservation. Of course, enticements that benefit individual organization members are simply bribery. This is one reason why many organizations try to avoid any perception of a conflict of interest and do not accept funding from businesses with a problematic reputation in environmental matters (e.g., marine conservation groups not accepting funding from oil and gas companies). It's a touchy issue, whereby maximum transparency can help avoid many unpleasant pitfalls.

Monbiot, G., Porritt, J. 2000. Does working with business compromise the environmentalist? *The Ecologist* 30(6): 20. Gale Academic OneFile.

https://go.gale.com/ps/i.do?p=AONE&u=anon~1fa85297&id=GALE%7CA65653617&v=2.1&it=r&sid=googleScholar&asid=4e014838

Robinson, J.G. 2012. Common and conflicting interests in the engagements between conservation organizations and corporations. *Conservation Biology* 26(6): 967–977. https://doi.org/10.1111/j.1523-1739.2012.01914.x

172. **You mean you want to prevent this major project and threaten all the potential jobs just because this small fish happens to live only here?**
- Small fish must always yield when the "big fish" wants to cash in mega.
- You have the power to decide whether a species will survive. Do you really want to abuse that power?
- Businesses and infrastructure projects can be set up almost anywhere, but some species may be restricted to the very spot you have in mind.
- Once again, the issue is one of balancing business interests with environmental integrity, ultimately the often-thankless duty of politicians (most of whom come from business, not from the field of environmental matters – so you have an inkling in which direction the decisions tend to go). The claim that environmentalists threaten jobs is a frequently heard and powerful allegation. It is also generally a half-truth or worse. The job number in question is typically a future projection and can be wildly inflated as a selling point for the project. Moreover, that dangled value might approach truthfulness only if the business thrives according to some highly optimistic business plan. Don't forget, predictions are a very difficult thing, especially if the future is involved. Importantly, the future personnel or employees may well be recruited from afar based on certain qualifications that might not be present in local residents (including yourself and your loved ones). Sometimes it helps to juxtapose the purported jobs promised by major projects (or the livelihoods threatened if that project is thwarted) to the jobs that leaving the environment intact can generate. Be that as it may, fighting any such project can expose conservation-minded folk to abuse ranging from vilification ("tree huggers," "communists," "fringe lunatics," "eco-terrorists") to death threats. Be cautious but stand tall!

By the way, the small fish alluded to earlier – the two-inch snail darter – was the first species whose restricted distribution and status under the US Endangered Species Act (ESA) was used to try to prevent the construction of a major project, the Tellico dam in Tennessee in the 1970s ("David versus Goliath"). The Supreme Court ruled in favor of the fish, but the Congress and the president then exempted the project from the ESA. Half a century later that fish proved to most likely be a subpopulation of another, more widely distributed darter species.

Ghezelayagh, A. et al. 2025. Comparative species delimitation of a biological conservation icon. *Current Biology* 35(2): 398–406. https://doi.org/10.1016/j.cub.2024.11.053

Plater, Z.J.B. 2004. Endangered species act lessons over 30 years, and the legacy of the snail darter, a small fish in a pork barrel. *Environmental Letters* 34: 289. https://heinonline.org/HOL/LandingPage?handle=hein.journals/envlnw34&div=16&id=&page=

173. **My business sector contributes only 10% to this pollution – why don't you go after the other miscreants?**
- Which of the many knife thrusts ultimately killed Caeser?
- Why shouldn't you take the lead and set a good example?
- Many types of pollution involve multiple sources. Every such source or sector contributes only a slice to the problem, and each likes to pass the responsibility on to another sector. Blame is the name of the game. Generally, that "other" sector is one that contributes a few percent more to the problem. Or it's one that may be viewed less benevolently by the public and therefore makes a more promising target for regulatory measures. The standard strategy is to try to deflect the pressure onto other sectors. The answer is not simply to compare, let's say, the emissions from cars to those of trucks, aviation, shipping, industry, or residential homes and then focus the blame or animosity on one of those other sources. Can this "blame" mindset be one of the steps in the historical development of your favorite country:

(1) land of the free, home of the brave (Ma, hand me the rifle),
(2) the ever-widening circle of blame (I'm gonna sue you),
(3) you hurt my feelings (get ready for the shitstorm)?

Of course, every sector must reduce its own footprint to the maximum extent possible, with pressure applied equally to all sources. It's the combined emissions that threaten the environment and, ultimately, human well-being and health. Shifting public opinion, perceptions, choices, and behavior patterns can play a key role in determining where public tolerance ends and which sector is targeted. Stricter and better enforced environmental legislation is also crucial. What's written in the lawbooks and what happens on the ground is one of *the* key problems facing every environmentalist around the globe.

Henrik Hammar, H., Löfgren, Å. 2010. Explaining adoption of end of pipe solutions and clean technologies – Determinants of firms' investments for reducing emissions to air in four sectors in Sweden. *Energy Policy* 38(7): 3644–3651. https://doi.org/10.1016/j.enpol.2010.02.041

Ritchie, H. et al. 2023. Emissions by sector. Our World in Data. https://ourworldindata.org/emissions-by-sector

174. **I pay my taxes – let the government fix this environmental problem.**
   - Passing the buck has never been the solution.
   - One US president put a sign reading "The buck stops here" on his desk. Unfortunately, that was apparently valid only for that president.
   - Don't forget, you *are* the government!
   - Yes, but you *are* the government (unless you happen to live in a dictatorship or perhaps in one of the more than two dozen "failed states" in this world)! One issue is what your taxes are being used for. Environmental challenges are only one of many problems that governments face (and may be low on their list of priorities). Are relevant environmental laws already in place? If yes, are they being heeded? Do you know what those laws say? Even tax laws, for example, can be written to provide incentives that promote conservation and biodiversity. Are you aware of them?

   Of course, personal initiative is always important in addressing any problem. Regarding the environment, it can start with you adopting the many "R's" in daily life (rethink, refuse, reduce, reuse, repair, recycle) and diligently separating your wastes (glass, paper, plastic, metal, etc.) in the respective bins. You can think globally (or nationally) but act locally. At the other end of your options, you can vote for the party that you think takes environmental issues most seriously. Have you ever considered writing to your local, regional or national elected officials? Or pushing for stricter laws? Or running for office yourself? Importantly, many environmental initiatives were first started by concerned individuals or citizen groups. Power to the people!

https://ici.gov.ck/wastemanagement/4rs-refuse-reuse-reduce-recycle/

Whitmarsch, L.E. et al. 2018. Waste reduction behaviors at home, at work, and on holiday: What influences behavioral consistency across contexts? *Frontiers in Psychology* 9: 2447. https://doi.org/10.3389/fpsyg.2018.02447

175. **I once donated to an environmental organization, so I've done my share.**
   - I'm sure that organization solved all the problems with your one-time donation.
   - "Giving is not just about making a donation. It is about making a difference." (Kathy Calvin)

- Nature conservation is a long-term effort. Every management project and program requires a long-term commitment in funds and personnel to sustain the results or until the system functions properly on its own. This means your effort should also be long-standing. Regular contributions, even if minimal, are often more important for longer-range planning than a one-time sum (unless you are bequeathing serious money). One engaged actor (Edward Norton) referred to the "white knuckle experience" of raising funds. Instead of conservation movements remaining donor-funded and philanthropic, he would reimagine the luxury ecotourism model to better support conservation efforts (for example in Africa).

But you can start simply. Pick an organization that best addresses your concerns. Or pick a local organization in which you might also consider becoming active in. You're a poor student? Ask your parents to become members. And the next time you are out in nature and encounter a team working on a conservation project (for example a sea turtle information booth at your beachside holiday destination), make a voluntary contribution. It's a "feel-good" proposition for both yourself and the project team. (For the remuneration of people working for environmental organizations, see entries Nr. 205 and 212.)

Verissimo, D. et al. 2018. Why do people donate to conservation? Insights from a 'real world' campaign. *PLoS One* 13(1): e0191888. https://journals.plos.org/plosone/article?id=10.1371/journal.pone.0191888

16
Politics

176. **This isn't your country. Why don't you go home and save your own animal species?**
  - I love your country as much as I hope you do.
  - Do you want to judge me by my passport or by what I have to offer?
  - I'm actually here because an international organization has recognized the environmental problem here as being a priority issue. The responsible authorities in your country are informed and have agreed to accept the help we can offer. We also have the specific training, expertise and funds to make a positive impact. And one part of our work is to train people from the local authorities/university/ministry here so that they can soon take over on their own. Of course, let us know if you or others have any suggestions, and let us know if anyone wants to help us out. I know we are guests, and we'll try to behave accordingly. We hope we can have the privilege of enjoying the hospitality your country is so famous for.

Chin, A. et al. 2019. The scientist abroad: Maximising research impact and effectiveness when working as a visiting scientist. *Biological Conservation* 238: 108231. https://doi.org/10.1016/j.biocon.2019.108231

Spiteri, A., Nepalz, S.K. 2006. Incentive-based conservation programs in developing countries: A review of some key issues and suggestions for improvements. *Environmental Management* 37: 1–14. https://doi.org/10.1007/s00267-004-0311-7

177. **You're from a landlocked country – why do you think we should trust you with solving the problems facing our sea?**
  - What's more important, my passport or my brain?
  - Astronomers don't live on the stars either.
  - My university has a top ranking in the world and we have a department that is specialized in marine science/conservation biology/ecosystem management. I have worked successfully in many other

DOI: 10.1201/9781003757474-16

countries and have been officially invited to work on this environmental issue here. Representing a landlocked country is also a good guarantee that I can tackle the work objectively, without any ulterior motives or hidden agenda. That's why landlocked countries are members of many international organizations and treaties dealing with the marine environment. The International Whaling Commission (IWC) is a convincing example. And you know, the Law of the Sea treaty (UNCLOS) actually recognizes that much of the world's marine resources are a common good of humanity and not restricted to those countries fortunate enough to have a shoreline. It's in all of our interests to keep the world's seas and oceans intact. Let's all cooperate to keep "our" oceans healthy. Together we are strong.

Sebuliba, S. 2024. The landlocked ocean: Landlocked states in BBNJ negotiations and the impact of fixed land-sea relations in global ocean governance. *Frontiers in Marine Science* 11: 1306386. https://doi.org/10.3389/fmars.2024.1306386

178. **You people in Europe/the United States etc. have destroyed your own environment, and now you want to tell us what to do?**

- Yes, and I'm hoping to help you avoid the same mistakes we made.
- You trust astronomers even if they don't come from another planet, don't you?
- In this increasingly interconnected world, no country has the sole responsibility for its environment and biodiversity. For example, many animals from birds to large terrestrial mammals to whales regularly cross national borders. Pollution also knows no borders, which is why research shouldn't be limited to the confines of national boundaries. This calls for international cooperation and coordinated action. That strategy is increasingly being recognized and facilitated by regional and global organizations, agreements, and conventions, most of them ratified by all the governments concerned, including yours. Foreign students and scientists play an important role in many countries. Most EU-funded projects, for example, seek a wide mix of participating countries both in and outside the EU. My cooperation partners in your country can confirm that we have strong financial backing and have developed a robust project with good chances of benefiting everyone. Of course, your country will be responsible for ensuring that this will be a lasting success even after we are no longer involved.

Rodríguez, J.-P. et al. 2007. Globalization of conservation: A view from the south. *Science* 317(5839): 755–756. https://www.science.org/doi/abs/10.1126/science.1

17
Radicalism

179. **You are a green extremist/eco-terrorist, why should we negotiate or compromise with you?**
- Look at my arguments, not at me.
- Don't shoot the messenger.
- "You must find a way to get in the way. You must find a way to get in trouble, good trouble, necessary trouble." (John Robert Lewis)
- Yes, we know that we are very quickly branded with not necessarily flattering names, whereby we can live with "do-gooder," "idealist," "naïve," or "optimist." And we certainly must be optimists because otherwise we wouldn't be here making this effort. The range of epithets, however, runs further from tree-hugger to eco-terrorist and worse (gasp, "communist"). How about simply agreeing on the term "activist," because becoming active is the only way you personally can help improve our world (see also entry Nr. 212). This all raises the important question of how, when, and by whom environmentally relevant issues are drawn to the attention of the general public or of the relevant authorities and decision- or policy-makers. In many cases, the problem is first recognized by researchers or by local stakeholders and then picked up by environmental groups. Many of the latter are anchored in widely recognized international or intergovernmental organizations. Of course, some groups are more activist, others more moderate. The basic idea behind all NGO strategies is to keep the pressure on. Maybe it's that mix of stated missions and operating strategies that promises the greatest progress and success in conservation-related issues. Importantly, decision-makers fear NGOs more than they do scientists (researchers don't tend to picket, start petitions, dump distasteful liquids in business offices, or rappel down buildings and tall smokestacks). Use that insight! How to get the message out there to the public and politicians? Clearly, the various organizations have different approaches. All should be based on

DOI: 10.1201/9781003757474-17

solid facts, which raises the crucial issue of bridging the so-called science-policy gap (see also entry Nr. 163). Accordingly, certain problems have been recognized for some time but no or insufficient action is being taken. This can be frustrating and helps explain why some organizations opt to take more dramatic steps. In a worst-case scenario, an environmental group may then be banned by a nervous or authoritarian government, and its top members may even be pursued and given jail sentences (for "illegal" activities or "plotting" against the government regarding protests they organized). We certainly hope that you don't approve of such an outcome in your country. In fact, environmental groups that have taken the initiative should be acknowledged and supported. Your family's well-being and future may depend on it.

Bondaroff, T.P. 2008. SECOND PRIZE: Throwing a Wrench into things: The strategy of radical environmentalism. *Journal of Military and Strategic Studies* 10(4). https://jmss.org/article/view/57662

180. **The Bible says we have dominion over animals and can do whatever we want with them.**

- God also gave us the gift of thinking straight. Maybe we should use it.
- The Bible also says that we should live after Jesus' role model. Destroying our planet isn't what He showed us.
- That's a tricky point and would require a deeper religious or philosophical discussion. Are we part of nature or separate from it? Are we accountable to something beyond humanity? Do animals have some intrinsic value beyond their utilitarian value to us? Do we have any responsibility to act in the best interests of future generations? The above Christian doctrine of "dominion," however, can be said to be based on the belief in a sharp division between human beings and nature, whereby the world would largely exist to serve human ends. Many people therefore treat the Earth as an automated teller machine (ATM) that hands us resources at our whim. And look where that position has gotten us.

One potential conclusion is that there is a certain orthodox religious arrogance toward nature, and that Christianity (but many other religions as well) bears a burden of guilt for the current environmental deterioration and over-exploitation of resources. Thankfully, many voices in religion have provided broader interpretations of those aspects of theological tradition that have been used to justify irresponsible attitudes toward nature (see for example the second encyclical of Pope Francis "Laudation Si"). Note also that, from a modern and practical ecological and environmental standpoint, humans are considered to be an integral part of nature, not separate from it (and therefore also need to be considered or accommodated for within conservation programs).

Harrison, P. 2006. Having Dominion: Genesis and the Mastery of Nature (1). *Environmental Stewardship*. London, T&T Clark International: 17–31. (Revised version of Harrison, P. 1999. Subduing the Earth: Genesis 1, Early Modern Science, and the Exploitation of Nature. *The Journal of Religion* 79(1): 86–109.) https://scholar.google.at/scholar?start=0&q=bible+dominion+over+nature&hl=de&as_sdt=0,5

https://iscast.org/opinion/why-should-christians-care-about-species-extinction/

181. **I can only warn you – if you keep up this conservation work here we will burn down your office/information booth/camp.**

- Please, let's talk about this over a cup of coffee or a beer.
- Maybe my kids can talk to your kids?
- We fully realize that conservation work can quickly ignite hostility from many sides. Anyone is rightfully skeptical when the younger generation or outsiders come in with new ideas and demands that might require local residents to deal with previously uncontested issues or even make changes in certain aspects of their lives. This, in turn, can make our fieldwork an unpleasant and risky business, so we'd like to sit down and hear your side of the story too. Tell us how we can do things better. First of all, let me say that we are not here out of a personal whim but rather because the government has given us approval and funded us to conduct this project. Moreover, what we are doing is fully in line with and backed up by national legislation and/or international agreements. The local authorities are fully informed of this program and have given us their backing. Of course, if we can successfully complete our project and protect this species/preserve or restore this habitat, then it will ultimately be advantageous for you and the entire region. After all, your environment is what makes this area so attractive in the first place. And you are proud of that, right?

Liu, T.M. 2017. Unexpected threat from conservation to endangered species: Reflections from the front-line staff on sea turtle conservation. *Journal of Environmental Planning and Management* 60(12): 2255–2271. https://doi.org/10.1080/09640568.2016.1273824

182. **You folks just aren't radical enough – why don't you burn down the offices of the people who are destroying our planet?**

- Live by the sword, die by the sword.
- As you sow, so shall you reap.
- Do unto others as you would have them do unto you.
- I'm afraid they all have surveillance cameras (and/or security staff).
- We love this planet, and that includes all its residents, even the misguided ones. The most important thing is to keep channels of discussion open. This means maintaining a certain status of competence and decorum. Mutual respect is a foundation for making progress on the negotiating table regarding conservation matters. That can quickly be destroyed by rash or extremist activity. That said, the pressure must be kept on those who are damaging our environment because, frankly, time is running out in many cases. The people at the negotiating table can include industry and tourism representatives, NGOs, IGOs, local residents, landowners, and indigenous peoples, along with a number of other stakeholders. A much more wide-ranging audience may also show direct interest because biodiversity, for example, is a global public good. Thankfully, a whole range of environmental groups with widely diverging tactics and agendas are out there, and their combined efforts – some perhaps more radical, some moderate – are probably the most promising approach to changing minds and practices.

Palmer, R., Timmermans, H., Fay, D. 2002. *From Conflict to Negotiation. Nature-Based Development on the South African Wild Coast.* Human Sciences Research Council, Pretoria. https://scholar.google.at/scholar?hl=de&as_sdt=0%2C5&q=conservation+negotiation&btnG=

18
Harebrained Ideas

Only two things are infinite, the universe and human stupidity, and I'm not sure about the former.

(Albert Einstein)

183. **Why shouldn't I make a campfire here? I'm an adult and know what I'm doing (Figure 27).**

Figure 27 Campfires, even near water, pose numerous risks. This is often compounded by the bad habit of trying to burn wastes rather than disposing of them properly. The technical terms for the products of such foolishness include pyroplastics, plastiglomerates, and plasticrusts.

- If adults always did everything right, we wouldn't need any laws.
- You can play your guitar in a circle without a fire.

DOI: 10.1201/9781003757474-18

- A cold can of beans has the same number of calories as a heated one.
- Careful, Smoky the Bear might just kick your butt.
- While making fire is one defining advance in the evolution of human societies, modern humans should restrict their fire-making to their barbecue contraptions on the patio or to designated fireplaces in the great outdoors. Why? Because Murphy's law always applies ("Anything that can go wrong will go wrong"). An estimated 90% of all wildfires globally are caused by humans. Of those, 70% are accidental (negligence, carelessness), 30% deliberate (arson). In the US, humans are responsible for 84% of all wildfires and for 97% of all those that threaten homes. High-resolution satellite data from hundreds of recent California wildfires shows that human-caused blazes spread much faster and kill more trees than those ignited by lightning. Other facts: humans extend the "normal" fire season, climate change is increasing the potential for wildfires, and post-fire soil and water degradation due to erosion is a key subsequent threat.

  On a practical, campfire level, the wood used to build and maintain a campfire will come from the immediate surroundings. Chopping down trees or removing that material from the ground is an interference in the habitat. Bringing your own firewood is also not recommended due to the risk of introducing harmful insects. One management strategy here is to ban axes, hatchets, and saws for hikers in the woods. Moreover, putting out a campfire is not a trivial task accomplished with a cup or two of water or a tiny camping shovel of dirt. Finally, most folks tend to try to burn some of their refuse in such fires rather than taking all the "icky garbage" back home. That more often than not includes tobacco products, food remains, plastic, aluminum foil, toilet articles, and glass. Only a small mistake, oversight, or a flying ember can lead to catastrophic forest fires that devastate entire regions and threaten human lives. So, if you do make a (legal) fire, read campfire safety tips first.

https://www.reserveamerica.com/articles/camping/top-10-tips-for-campfire-safety/

https://www.science.org/content/article/human-sparked-wildfires-are-more-destructive-those-caused-nature

Reid, S.E., Marion, J.L. 2005. A comparison of campfire impacts and policies in seven protected areas. *Environmental Management* 36: 48–58. https://doi.org/10.1007/s00267-003-0215-y

184. **I may be feeling awfully sick, but I'm still going on today's tour to see the gorillas, considering the amount of money I spent!**
   - If your country spent lots of money on nuclear weapons, would you argue for using them to make it worthwhile?

- If you were highly contagious, would you go visit your grandparents or grandchild?
- You'll be doing neither yourself nor the gorillas a service by going on that tour. Think of your own safety: if the exertion suddenly worsens your condition, do you really want to be treated in the nearest village clinic? Do you really want the whole group to have to turn back because you had to give up? The major conservation concern, however, is the potential transmission of your illness. Gorillas are among our closest relatives, which means that they are susceptible to many of the same diseases we have – and may have immune systems that are less efficient against the respective pathogen. This is an unacceptable risk for members of this (or any) endangered species. Finally, every individual animal counts. In a worst-case scenario, you could be responsible for infecting a key individual or even wiping out an entire gorilla family. Unfortunately, despite common sense and pre-trek briefings, even the simplest rules, such as keeping minimum distances (e.g., 7 m) tend to be broken in virtually all tours, leading to discussion about tourists wearing face-masks to protect gorilla health.

Weber, A. et al. 2020. Lack of rule-adherence during mountain gorilla tourism encounters in Bwindi Impenetrable National Park, Uganda, places gorillas at risk from human disease. *Frontiers in Public Health* 8. https://doi.org/10.3389/fpubh.2020.00001

185. **When and where can I pet the bears?**
- Hmm, there seems to be a significant overlap between the smartest bear and the dumbest tourist.
- Never underestimate an animal that can be trained to ride a bicycle.
- Is that crunching sound your cell phone or your finger/arm bones?
- I'm afraid you don't have the correct impression about wildlife. As the first half of the word states, they are wild (not tame or pets, whereby even that can't guarantee your safety). And a GPS-collared bear is not tame either. It's hard to think of any wild animal that can't pose a threat to a human. Simply put, a bear can tear you apart in seconds. More broadly, bears wander through some national parks (and elsewhere) at will, and it isn't possible to predict where they might be at any time (and how indisposed they are: for example, you know how cantankerous you are with a toothache, now translate that to a top predator with a toothache . . .). Second, most parks specify mandatory minimum distances, and for bears, that's somewhere around 100 yards/meters, that is, basically the length of a soccer/football field. In bear country, most people go to great efforts to keep them at a distance (a well-trained dog often does the trick) – why should you be

seeking a close encounter? Finally, repeated encounters with humans can alter bear behavior (habituation). Feeding them to get closer or attempting a selfie is a good way to turn them into so-called "nuisance" bears, many of which must ultimately be shot. If you're in your car and they get closer, roll up your windows to avoid potentially painful (and/or expensive) misunderstandings!

Ordiz, A. et al. 2013. Lasting behavioural responses of brown bears to experimental encounters with humans. *Journal of Applied Ecology* 50(2): 306–314. https://doi.org/10.1111/1365-2664.12047

186. **The whale had fishing gear wrapped around it, so I jumped into the water with a knife to free it.**
    - Would you jump on an entangled elephant with a knife (well, at least you wouldn't drown . . .)?
    - Over 300,000 whales and dolphins die annually due to entanglement in fishing gear. This is a serious welfare issue because such animals tend to die a slow and painful death. The International Whaling Commission (IWC) has therefore launched a Global Whale Entanglement Response Network. This involves an expert panel, international training courses, the development of special cutting tools, and "best practice guidelines." The take-home message of the latter is that whale disentanglement must be left to the professionals who follow strict protocols: no boater, local resident, or tourist should ever enter the water to try to free an entangled cetacean on his or her own. Even trained teams are not always successful. Human safety is the number one priority, and uninitiated individuals have been killed in such attempts. An injured and often panicked whale – not to mention loose ropes and large nets in the water – is a recipe for disaster. Boaters should have the telephone number of the local entanglement response team readily at hand.

    https://iwc.int/management-and-conservation/entanglement/entanglement-response-network

187. **If you carefully slip your hand underneath a stonefish and lift it up, it can't harm you.**
    - Yes, and you can catch a rattlesnake with your bare hands if you are talented and lucky.
    - Who has the bigger brain: you or the fish?
    - Never trust an animal with a small brain. (Mike Stahovich)
    - Stonefish are the most venomous fish in the world, and a sting from their dorsal spines is extremely painful and potentially fatal (apply heat and hope an anti-venom is available). Most incidents involve

stepping on them in shallow water (their spines can penetrate a soft shoe sole). They are often buried in the sand with only eyes and mouth slightly protruding, making them visible only upon closer inspection with a trained eye. Eager but imprudent dive guides have been known to scoop them out to impress their group. It's a classic case of harassment, machoism, and simply foolishness. And that harassment may continue for a considerable time as divers line up to take photos. This is yet another case where even taking a photo can be an inappropriate behavior that shows disrespect for wildlife in their natural setting.

Connors, N., Nelson, L.S. 2013. Fish "sticks". *Emergency Medicine* 45(2): 9–21.

Lyon, M.R. 2004. Stonefish poisoning. *Wilderness & Environmental Medicine* 15(4): 284–288. https://doi.org/10.1580/1080-6032(2004)015[0284:SP]2.0.CO;2

188. It was only a harmless nurse shark, so I snorkeled down and grabbed its tail while it was lying on the bottom (Figure 28).

Figure 28 Only one of our five senses (touch) works normally underwater. This might help explain our urge to pat, pet, grab, or otherwise make physical contact with marine organisms. Glove bans (see Entry Nr. 87) can help counter that but even gloves wouldn't protect you from a bite by a hungry and agitated nurse shark.

- You look harmless, too. Would you like it if I grabbed your tail?
- Any wild animal bigger than a chihuahua (and some much smaller) can really hurt you (or your kid, or your baby).
- While nurse sharks have only very small (but sharp and many) teeth that will not cause major wounds, they have been known to bite people in self-defense or upon being otherwise harassed (for example by grabbing their tail) or while being fed. Why someone would even think of grabbing a fish that can reach 3 m in length and weigh over 100 kg is yet another sad commentary on our disconnect with and poor understanding of wildlife. Nurse sharks are considered to be "docile" but nonetheless rank quite high in the number of "attacks" on humans. They may clamp down tightly on their victim, and some do not let go. You will no doubt have difficulty swimming to the surface (or remaining on the surface) with a large, heavy, angry, thrashing fish attached to you and heading for deeper water, so the question is how long can you hold your breath in such a panic situation. As is the case for many shark species, nurse sharks are threatened in some areas. Never pet, grab, feed, or take an up-close selfie with any wild animal.

Boissonneault, M.-F. et al. 2005. Grey nurse shark human interactions and portrayals: A study of newspaper portrayals of the grey nurse shark from 1969–2003. *Electronic Green Journal* 1(22). https://escholarship.org/uc/item/9nb9h48n. https://doi.org/10.5070/G312210610

https://ioa.factsanddetails.com/article/entry-187.html#chapter-10

189. **In its native country that species eats an important pest, so we're introducing it here to eradicate a similar pest.**
   - Define pest.
   - That strategy has never ever worked before, so why here and now?
   - Don't be surprised if they ultimately take over your backyard too.
   - If you're lucky, it tastes great, and the necessary eradication program may put meat on your table.
   - We have to face the fact that almost every country has problems with so-called pests (see entry Nr. 8). Some are native species, others are introduced aliens. Fighting any pest species by introducing a predator of that or a similar species elsewhere is tantamount to pouring oil on the fire – combating one evil with another. Sometimes the notion is sold under the catchy heading "biological pest control," which may sound better than chemical control. While the biological pest control approach encompasses many promising techniques, all alarm bells should ring when one animal is introduced to combat another.

The well-known results: the introduced predator may target and harm the populations of entirely different and desirable native species, and the population of the introduced species can explode and become highly invasive, sometimes requiring time-consuming and expensive control measures. Unfortunately, repeatedly documenting new cases or explaining why this occurs (e.g., "enemy release hypothesis") helps little to rectify the situation. Australia is frequently mentioned as a classic case, with some of the aliens having been deliberately introduced to fight other species (e.g., cane toads to fight cane beetles; myna birds to control insects; mosquito fish to control mosquitoes), with significant impacts on the environment, economy, and human communities.

Roy, H.E. et al. 2011. Can the enemy release hypothesis explain the success of invasive alien predators and parasitoids? *BioControl* 56: 451–468. https://doi.org/10.1007/s10526-011-9349-7

West, P. 2018. *Guide to Introduced Pest Animals of Australia*. CSIRO Publishing. https://www.publish.csiro.au/book/7538/

190. **We'll fertilize the oceans with iron to make them work better!**

- Define "better."
- It's not nice to fool Mother Nature.
- And while we're at it, let's divert a few rivers to more arid regions, block the sun's rays with space umbrellas and release sulfur dioxide into the stratosphere (to cool the Earth), and simply store all that evil CO_2 underground, etc.
- Like many megalomaniac ideas (often under the heading of "geo-engineering"), fertilizing certain ocean regions to help "improve" their productivity or help fight climate change was briefly a hot topic that is largely discredited today. The more we begin to understand ocean dynamics along with both microscopic and large-scale ecosystem interrelationships, the less sway such notions have. Importantly, the over-fertilization or eutrophication of many coastal areas has been identified as a key marine pollution problem – in fact the only form of pollution that can kill off entire marine environments (leading to so-called "dead zones"). Such "cultural or anthropogenic eutrophication" is, as the name implies, an entirely human-driven problem. Why, then, we would willingly wish to fertilize other oceanic areas remains a mystery of twisted human logic. It's tantamount to remediating one pollutant by introducing a second pollutant. Maybe some folks hope to make those regions more "productive" (yield more fish?). All these desires are certainly not a reason to try to turbocharge the system. If we can't extract more fish from the sea despite ever-greater efforts (and that is the case), then that's the direct definition of over-fishing,

that is, yet another conundrum we have created ourselves. As Shakespeare more eloquently wrote, "The fault is not in our stars but in ourselves." We would be better off tempering our demands than further stoking the (marine) environment or seeking blame elsewhere.

Hale, B., Dilling, L. 2020. Geoengineering, ocean fertilization, and the problem of permissible pollution. *In: The Ethics of Nanotechnology, Geoengineering, and Clean Energy* (pp. 285–307). Routledge.

191. **It's OK to release this exotic animal because I don't think it can breed here/I think we can prevent it from reproducing.**
    - That's what the owners of "Jurassic Park" thought too about their dinosaurs.
    - "I don't know the question, but sex is definitely the answer." (Woody Allen)
    - ". . . in nature, procreation is the single relentless rule." (Camille Paglia)
    - Almost nothing can stop animals (or humans, for that matter) from reproducing. The reproductive mechanisms in the animal kingdom are so manifold, and the drive to reproduce is so strong and evolutionarily anchored, that believing it can be prevented is simple lunacy. The notion of entirely thwarting reproduction merely by putting those animals in a different environment, or by somehow constraining or confining them (hey, we are introducing them only into this one lake!) is ill-advised at best. The same holds true for applying some form of sterilization, or by using contraceptives and other hormonal treatments. Some of the introduced individuals may be pregnant females. Other species don't even need males to reproduce. Yet others are initially males, then females when large, or the other way around (protandry/protogyny). Finally, even a released non-reproductive animal may be long-lived and have years or decades to wreak havoc in the wild.

Hampton, J.O. et al. 2015. Is wildlife fertility control always humane? *Animals* 5: 1047–1071. https://doi.org/10.3390/ani5040398

Massei, G. 2023. Fertility control for wildlife: A European perspective. *Animals (Basel)* 13(3): 428. https://doi.org/10.3390/ani13030428

192. **The freshly landed shark/shot rhino, lion etc. stopped moving, so I got up close for a good selfie.**
    - The way you hold that shark may well determine which organ you might lose.
    - You Tarzan?
    - (Your) Jane is probably not enthusiastic.

- Hunting is an integral part of human history and helps explain our survival and great success as a species. Today, it (big-game hunting at least) is more often than not a "sport," which could be interpreted as being positive in the sense that it may be one of the only contacts we have with some semblance of intact nature. Even that argument loses some of its persuasiveness if the animals have been bred and released (and sometimes apparently lightly sedated) shortly before being shot. It can also be argued that some animals must be hunted to control their populations (often, however, only because we have "unbalanced nature" by eliminating their predators in the first place). Unfortunately, the negative aspects – ranging from animal welfare considerations to the role of "bush meat" in further reducing the populations of endangered wildlife – tend to outweigh any benefits for most folks. Of course, hunting with a camera instead of a rifle, that is, not killing the wildlife, was long an ethical alternative. Today, however, selfies (in the framework of "selfie safaris") with living wild animals have also been recognized as an animal welfare issue and as being potentially harmful. What the exact message you wish to convey by uploading photos of yourself up close with a freshly killed trophy is unclear. You may find that lots of folks end up seeing that image when it ultimately ends up on social media & Co. (What!? I only sent it to a few close friends!) And most people these days – maybe your prospective employer, or the guy or gal you wish to have a meaningful relationship with, or the nature conservation organization that asked you to be their figurehead (I'm not kidding . . . King Juan Carlos of Spain) – may not necessarily be amused.

Kalof, L., Fitzgerald, A. 2003. Reading the trophy: Exploring the display of dead animals in hunting magazines. *Visual Studies* 18(2): 112–122. https://doi.org/10.1080/14725860310001631985

Medley, C. 2024. Selfie safaris: The violence of contemporary camera hunting and trophy shot selfies. *In: Violence and Harm in the Animal Industrial Complex* (pp. 113–133). Routledge.

193. **I'm not allowed to smoke indoors, so now I smoke a lot in the great outdoors.**
- An addict is an addict.
- And I'm sure you take all your cigarette butts home with you?
- If you smoked at home at least you couldn't start a forest fire.
- I'm sorry to hear that you feel you still need to smoke. But I understand what addiction means. Did you know that some forest firefighters actually use their breaks to smoke a cigarette? Interestingly, secondhand smoke is also an issue in outdoor environments: it's considered to be a public health concern because there is no safe level of exposure. As you

probably know, cigarette smoke contains hundreds of chemical compounds, many of them toxic. Unfortunately, that makes you a walking source of air pollution. Although the concentration of the smoke is no doubt lower in open spaces, you'll seldom be the lone smoker (and seldom be alone). And, of course, after a smoke, each cigarette butt is a tiny toxic bomb (don't forget, the filter itself is largely plastic, not paper). Hope you're taking them all back home for proper disposal.

Fiore, M.C. et al. 1990. Methods used to quit smoking in the United States: Do cessation programs help? *Journal of the American Medical Association (JAMA)* 263(20): 2760–2765. https://doi.org/10.1001/jama.1990.03440200064024

US Department of Health and Human Services. 2006. *The Health Consequences of Involuntary Exposure to Tobacco Smoke: A Report of the Surgeon General.* US Department of Health and Human Services, Centers for Disease Control and Prevention." Coordinating Center for Health Promotion, National Center for Chronic Disease Prevention and Health Promotion, Office on Smoking and Health.

194. **Animals and I are on the same wavelength. It's a special gift I have, a karma thing, they accept me. I'll give it a bit of my sandwich and then try to cuddle it.**
    - I'm sure your cheese sandwich will be a balanced enrichment of its diet.
    - I'm sure the selfie will show where you are bleeding the most.
    - Do you have an organ donor card?
    - You might be in for a major surprise in trying to get up close to wildlife (and as a matter of fact that can apply equally to unfamiliar pets). Of course, if you get close enough, you'll no doubt also want to take a selfie of your encounter. And don't forget that you probably aren't the only one with this brilliant idea. Is there a line building up behind you? Wildlife that regularly receives tourist-provided food (or even merely attention) can become dependent on and habituated to human contact, which is rarely positive for either side. Are you sure the animal can distinguish between the food and the wrapper? Or whether it can or wants to distinguish between your fingers and the food? Maybe the animal is big enough to swallow both whole? Moreover, typically more than one animal will be attracted, which can lead to fights that harm not only the animals but also bystanders. Another factor to consider when getting up close to wildlife is the potential transmission of diseases (in both directions!). Finally, let's mention "trust." And I'm not referring to trust in animals (although how much trust should you place in animals with considerably smaller brains than yours?). Do you trust

your dive guide if he or she says it's OK to pet the moray eel, or the whale watching operator who says it's OK to jump in and swim with the orcas, or the trainer at the marine theme park who urges you as a spectator to put your arm around a seal and let it "kiss" you? If things go wrong, they may get fired, but you get to keep the scars.

Cater, C. 2010. Any closer and you'd be lunch! Interspecies interactions as nature tourism at marine aquaria. *Journal of Ecotourism* 9(2): 133–148. https://doi.org/10.1080/14724040903125039

Orams, M.B. 2002. Feeding wildlife as a tourism attraction: A review of issues and impacts. *Tourism Management* 23(3): 281–293. https://doi.org/10.1016/S0261-5177(01)00080-2

195. That dog/coati/racoon/you name it/stole my sandwich. I'll teach it a lesson! (Figure 29).

Figure 29 Many people today have lost their understanding of and respect for wildlife. When wildlife loses its respect for people, things can get unpleasant fast. Intervening in this encounter risks losing blood.

- That's one food fight you'll probably lose.
- Oh my, that gash will require at least 10 stitches.
- Better your sandwich than 3 fingers.
- It *was* your sandwich. Now it's *his/her* sandwich. If you think you're mad at losing it, get ready for a major escalation in trying to recover it or otherwise threatening the animal. When do most animals show

aggressive behavior? Fighting for mates, territory, and food are no doubt at the top, whereby food is probably the most frequent one ("resource guarding"). An animal's rank among its kin determines its willingness to and success in engaging in fights (intraspecific competition). And within a species, two contestants typically stop before major injuries occur and use subtle behavioral signals to indicate submission. For those outside that clique (and you yourself are a true outsider), the chances are good that it will be no-holds-barred contest. Every dog owner knows not to take food away from their pet with their bare hands and fingers. Apply that wisdom to animals who have never lived in a house. Forfeit the food and chalk it up to experience. It's a food fight you can't win.

Jacobs, J.A. et al. 2018. Defining and clarifying the terms canine possessive aggression and resource guarding: A study of expert opinion. *Frontiers in Veterinary Science* 5: 115. https://doi.org/10.3389/fvets.2018.00115

196. If you grab a sea turtle underwater, it will take you on a wonderful ride (Figure 30).

Figure 30 What wildlife is suitable for trying to take a ride on? None, of course! Why should a sea turtle be any different? Always maintain a respectful distance (and please no selfies!), even if the species stays put and doesn't flee. And don't forget, sea turtles have very powerful jaws and can bite! (Image from Shutterstock).

- How would you like it if a big guy jumped on you for a piggy-back ride?
- Hey, that sounds like a great idea – the nature equivalent of subway surfing.
- You may think no one would ever come up with such a harebrained idea, but many (mostly scuba divers or brash snorkelers) have. This shows a total disconnect with nature and blatant disregard and disrespect for wildlife. Holding an air-breathing animal underwater is certainly a severe threat, and the "ride" can be a panic reaction that requires the turtle to waste valuable energy. Moreover, its activity at the time (sleeping, feeding) will have been interrupted. The internet also contains photos (thankfully, mostly "historical") of people sitting on or riding sea turtles on the beach. Turtle crawls on the beach are laborious enough as it is, and sitting on them adds insult to injury. It can also disrupt nest laying, a short but crucial time in a sea turtle's life cycle. Such harassment is also illegal (a felony in some US states) and is heavily fined (up to $10,000 for first offenders under the Endangered Species Act; up to five years in prison and/or a fine of up to $5,000 if violating the Marine Turtle Protection Act in Florida). Putting such a photo on social media, of course, proves the stupidity of the miscreants (and can be used against them in court).

https://www.dnr.sc.gov/marine/pub/seascience/seaturtle.html

http://www.hawaiioceanambassadors.com/beach-clean-ups/info-sea-turtles-can-i-touch-or-ride-them

19
Pessimism/Nihilism/Cynicism

197. **In 200 million years, the solar system is going to collapse anyway, so why worry about a few species now?**
   - "It's the end of the world as we know it and I feel fine." (R.E.M.)
   - We have difficulty thinking ahead for even ten years, so let's concentrate on improving that ability and on tackling the current reality rather than getting into cosmic dimensions.
   - Predictions are a very difficult thing, especially when the future is involved.
   - Maybe our world really will come to an inglorious end eons from now, but you need to ask, based on the current state and rate of decline, how soon and what kind of devastation your kids and grandkids will be facing in the more immediate future. And, of course, life is mostly about the here-and-now, so how will all this already affect you personally in the upcoming years? Regardless of when you were born, several of the recent human generations have also been given the "Me Generation" label. That means it's not just about the fate of a few species or biodiversity in general. It's also about the quality of the water *you* drink, the food you eat, and the air you breathe. That means it's about *your* fate as well. And intact nature is actually what delivers all those services and thus helps decide your future. And as far as the air is concerned, don't forget, if you can't breathe, nothing else matters.

Besley, T., Peters, M.A. 2023. Life and death in the Anthropocene: Educating for survival amid climate and ecosystem changes and potential civilisation collapse. *In: Educational Philosophy and Post-apocalyptical Survival* (pp. 60–75). Routledge. https://www.taylorfrancis.com/chapters/edit/10.4324/9781003453895-7/life-death-anthropocene-tina-besley-michael-peters

DOI: 10.1201/9781003757474-19

198. **I'm not going to be around in the future anyway, so why shouldn't I enjoy life to the full without bothering about conservation issues?**
- "Don't worry, be happy." (Bobby McFerrin)
- That sounds pretty selfish to me.
- Think that through to the end: you'll probably be dead within the next few decades or so, so why not just buy a good TV and binge watch until it's game over.
- Almost every religion teaches that it is not only about us, but about our fellow human beings too.
- Go for it. After all, even the US Declaration of Independence recognizes the unalienable right to "the pursuit of happiness." But we can ask what truly ensures that happiness and a healthy life. The most basic answer may well be an environment that is sufficiently intact and functioning to supply us with the food and water we need and to provide the aesthetically and psychologically edifying landscapes we need. Countless studies and common sense underline that tall buildings, asphalted surfaces, dense traffic, digital worlds, and the like cannot meaningfully provide that service. And you don't have to look far to hear that money won't necessarily make you happier either. It's a sad state that we have become powerful enough to change the face of our planet but are loath to modify our own behavior to ensure that it remains livable. It's a potential tragedy, and you don't really want to be among those who fail to recognize that, do you? Of course, if you or your family have children and grandchildren, you might want to consider what kind of environment you hope to leave behind for them to be happy in. So why don't you step up and do your share now?

Hamilton, C. 2017. *Defiant Earth: The Fate of Humans in the Anthropocene.* John Wiley & Sons. https://scholar.google.at/scholar?hl=de&as_sdt=0%2C5&q=Earth%27s+fate&btnG=

MacKerron, G., Mourato, S. 2013. Happiness is greater in natural environments. *Global Environmental Change* 23(5): 992–1000. https://doi.org/10.1016/j.gloenvcha.2013.03.010.

199. **We'll have to rely on our children's generation to solve this mess.**
- So, you want to pass our sins off to the next generation?
- "You cannot escape the responsibility of tomorrow by evading it today." (Abraham Lincoln)
- I'm sure your kids will love and honor you for shirking your duty.
- It is true that many conservationists place their hopes in the next generation. This stems from two lines of thought. Firstly, the recognition that the opinions and practices of many of today's "older"

generation cannot be changed sufficiently to bring about true change fast enough. Secondly, the expectation is that the education of the younger generation will address environmental issues that earlier generations never heard of or had to deal with. Understanding children's attitudes toward the environment can also help shape current conservation approaches. In the meantime, many hope (and see) that youngsters often shame their elders into more environmentally friendly behavior ("What? I can't believe you want to leave all our picnic garbage on the beach!"). And, of course, it's always a good strategy in any confrontation to mention children – yours, your discussion partner's, all future generations – if you want to enlarge the context and send your message on a more personal level.

Battisti, C. et al. 2018. Children as drivers of change: The operational support of young generations to conservation practices. *Environmental Practice* 20(4): 129–135. https://doi.org/10.1080/14660466.2018.1541679

Shapiro, H.G. et al. 2016. Which species to conserve: evaluating children's species-based conservation priorities. *Biodiversity and Conservation* 25: 539–553. https://doi.org/10.1007/s10531-016-1067-0

200. **We'll only solve this mess if the human race gets wiped out.**
- Theoretically yes, but some of us will probably survive no matter what kind of maniacal folly we might stoop to.
- Humans are a quite hardy bunch, so wiping us out would probably wipe out most other higher forms of life on the planet too.
- In earlier times, environmental scientists thought of the natural world as being that before the advent of humans and our many impacts (sometimes referred to as the pre-exploitation state or historical baseline). That meant factoring people out of nature conservation efforts. According to that notion, national parks and other protected areas were to be established without any human presence or influences, in some cases forcibly resettling people within park boundaries elsewhere. This has changed. The modern, more holistic perspective is to consider humankind as an integral part of nature today (e.g., the One Health approach, see entry Nr. 203). Future plans should therefore include the human component. Of course, the challenge is to ensure that we all do our share in keeping our planet livable – and thus prevent ourselves from being wiped out.

Kantor, I. 2007. Ethnic cleansing and America's creation of National Parks. *28 Public Land & Resources Law Review* 41. https://scholarworks.umt.edu/plrlr/vol28/iss1/

Kurle, C.M. et al. 2023. Considering humans as integral components of "nature". *Ecological Solutions and Evidence* 4(1). https://doi.org/10.1002/2688-8319.12220

201. **It's hopeless. We as individuals are powerless to stop environmental degradation.**
    - Does that mean you think you are "only pawn in game of life"? (Blazing Saddles)
    - I'll accept that excuse if you are comatose.
    - In fact, each one of us can make a difference. It can start with your own behavior and lifestyle. The next step is being a role model for your family and friends. To quote the anthropologist Margaret Mead: "Never doubt that a small group of thoughtful committed individuals can change the world. In fact, it's the only thing that ever has." A keyword there is "committed." How much do you care? Enough to seek out others who might join you? Enough to take action together. What might that action be? If you are a student, organize campus activities. If you are a teacher, embed the message into your lectures. Engage in your professional networks. If you are an artist, address the issue in your artwork. Importantly, get in touch with your elected officials from the local to the national level. Or how about running for office yourself?

      And, of course, money talks: buy environmentally friendly products and boycott those that aren't. If you're a bit wealthier, you can try to influence corporate behavior and consider withdrawing investments from companies with poor environmental policies. And if you are even wealthier, who are you considering bequeathing your wealth to? The founder of the outdoor company Patagonia gave away 98% of his three-billion-dollar business to a non-profit that advocates environmental causes (those funds were recently used to enable conservationists to buy the site of a planned titanium mine at the edge of Georgia's Okefenokee Swamp). In an allied endeavor, an American couple purchased more than two million acres of land in Chile and Argentina, created a foundation, and then, in an ongoing effort, donated the land to the governments to help establish or expand 14 national parks. Wow.

      From another perspective, we don't have the luxury of being hopeless or pessimistic: "If you don't have hope that your action is going to make a difference, why bother to do anything? You just become a zombie." (Jane Goodall). Yes, some of the best people are giving (or gave) it their all. And there are other grounds for optimism! So-called "lost species," that is, those that haven't been seen for at least a decade and could be extinct, are sometimes resighted (e.g., the Madagascar pochard – a duck). Sometimes we luck out and can bring a species back from the brink (e.g., the Florida grasshopper sparrow, American bison/buffalo). Captive breeding (for example of animals that are extinct in the wild) and reintroducing them into their former habitat can also be a new lease on life (black-footed ferret, scimitar-horned oryx). The hybrids of one of the three extinct species of Galapagos tortoises, for example, are slated to be returned to their native Galapagos island soon (after getting tough on the feral pigs, rats, dogs, cats, ants, and other threats there). Numerous other species have been "delisted," that is, their threat status improved a notch. Dedicated nature conservation and species protection workers are behind all such success stories.

How we saved the world. An optimist's guide to life on Earth in 2070. How we lost the planet. A pessimist's guide to life on Earth in 2070. National Geographic 04.2020.

Kenis, A., Mathijs, E. 2011. Beyond individual behaviour change: the role of power, knowledge and strategy in tackling climate change. *Environmental Education Research* 18(1): 45–65. https://doi.org/10.1080/13504622.2011.576315

202. **We'll have to wait for some future technology to solve this environmental problem.**

- Technology is the answer. But what was the question?
- Maybe, but what about simply using common sense now?
- High-tech never solved any problem for a lowlife.
- Living within ecological boundaries is not optional. (Peter Frankopan)
- Yes, one thing we can be sure of is that technology will advance at an ever-increasing clip. Unfortunately, as experience shows, most of the problems we face today are also ballooning at a prodigious rate. And many problems were created by technology in the first place. On the one hand, technology always seems to be one step behind. On the other hand, each new technology seems to create its own new set of problems.

In the conservation field, the good news is that practitioners are typically avid consumers of technology to quantify and help solve the issues they face. Moreover, many are also technology innovation leaders whose labs, workshops, and industry collaborations are busy developing new devices to go deeper or higher or longer, to measure ever more accurately, or to further miniaturize their instruments for new nature conservation challenges. The bad news is that the major issues, such as habitat loss, species extirpations, and the establishment of protected areas, are more a matter of human action and political will than of current technological shortcomings. Waiting for future generations and novel technologies cannot be the answer, considering the rapid pace of environmental decline. Extinctions, for example, are forever, and no future technology will bring a lost species back (even though some may be discussing it; see "de-extinction" under entry Nr. 26). Rather, we must use common sense and promote beneficial policies in the here and now. And don't forget, no matter how sophisticated our technologies may become, we'll always be just a week or two without electricity from starting to burn our furniture for heat and reverting to a more brutal hunter-gatherer status.

Berger-Tal, O., Lahoz-Monfort, J.J. 2018. Conservation technology: The next generation. *Conservation Letters* 11(6): e12458. https://doi.org/10.1111/conl.12458

Wich, S.A., Piel, A.K. 2021. *Conservation Technology*. Oxford University Press. https://doi.org/10.1093/oso/9780198850243.001.0001

203. **As long as I'm healthy, why should I care about nature conservation or species protection?**

- That's taking the "me" generation one step too far.
- "I me mine." (George Harrison).
- One key insight being gained in many disciplines today is *interrelatedness* (even if many scientists still get flak for deviating into interdisciplinary research). This goes far beyond the environmental sciences to include purely human pursuits such as political science and economics. Everything is interconnected. The result in environmental sciences has been a paradigm shift in which humans are considered to be an integral part of nature. Accordingly, our fates are intertwined. This perspective is at the core of the One Health approach, as advanced by the World Health Organization (WHO). This approach recognizes that "the health of humans, domestic and wild animals, plants, and the wider environment (including ecosystems) are closely linked and interdependent." Human activities and stressed ecosystems have created new opportunities for diseases to emerge and spread, whereby such stress factors can involve animal trade, agriculture, livestock farming, urbanization, extractive industries, and climate change, as well as increasing habitat fragmentation and our ever-further encroachment into wild areas. This perspective focuses primarily on the spread of infectious diseases from both wild and domestic animals, which has become a major human health issue. And, of course,

infectious diseases are a two-way street between animals and humans (see also entry Nr. 184). The call is therefore to sustainably balance and optimize the health of people, animals, and ecosystems. In short, and quite logically, your health (and human health and survival in general) cannot be ensured if the environment goes to hell. The One Health approach is being considered by a wide range of fora, including, for example, the International Whaling Commission (IWC), where it has been advanced as one potential framework in the IWC's Scientific Committee Environmental Concerns Sub-Committee.

https://www.who.int/news-room/fact-sheets/detail/one-health

https://iwc.int/management-and-conservation/environment/cetacean-diseases/infectious-disease

https://www.fao.org/animal-health/animal-diseases/highly-pathogenic-avian-influenza/en

204. **I don't like your tough language – these organisms are here for our own use, for our very survival!**

- The line between use and abuse is very thin.
- Watch your language! (every mother's admonition).
- Your take on survival contains an element of truth. Altogether, functioning ecosystems along with their inhabitants *have* always been used by us and *do* help guarantee our survival. Conversely, our chances of survival would be low – and our well-being severely compromised – should nature be reduced to an artificial event backdrop. This means not overexploiting living natural resources to the point of local extirpation or extinction. Why has this become an issue? The increasing global human population, ever-improving hunting and fishing technologies, and in some cases traditional religious interpretations (see entry Nr. 180) play a role.

But language also plays a role. It reflects and shapes our attitudes and actions. The terminology used to describe our relationship with wildlife and nature is often euphemistic or tailored to human use. Such language has been introduced because, deep down inside, most people are queasy about overexploiting nature in an unsustainable manner, and we seek to linguistically unburden or circumvent personal sensitivities. Accordingly, we use neutral or more palatable terms such as "stock" when referring to some part of an animal population that we have set our sights on, or refer to "take" and "harvest" when we actually simply mean "kill." Or "depopulation" when we "cull" millions of birds in highly disconcerting procedures during avian influenza events. Even the term "quota" implies an acceptable and harmless "removal" from a natural population – which has led to disastrous results in the past and continues to be a heated, bitter issue between fishers, scientists, and politicians. Another typical example is the term "sacrifice," often used by researchers to describe the killing of laboratory animals.

Such terminology has lulled many to believe that how we treat nature or what we take from it is entirely admissible. Using language to help take personal sensitivities into account is one matter. Being purposefully manipulative is another, for example, when euphemisms are used to mislead or deceive (e.g., "doublespeak) or when lobbyists apply language that avoids (or promotes) an emotionalization of the public discourse, depending on the interests they represent. Such terms were often coined many decades or even centuries ago. Our improved understanding of animal behavior and cognition, their roles in the ecosystems, along with a new appreciation of animal rights and (bio-) ethics, has sparked a discussion on finding more accurate and suitable alternatives. Adopting a new language may be a step forward to a more respectful understanding and management of wildlife. Let's tell it like it is.

Nurse, A. 2020. Value-Laden Language and its consequences for wild animal welfare. A report for the Wild Animal Welfare Committee. https://static1.squarespace.com/static/5edf4fd72d25275e3acc8c4a/t/5f5258d0ed0dbd7b5193f25e/1599232210487/Topic+Statement+Language+Final+Sept+2020.pdf

Ritter, F. et al. (submitted). Words matter: A call to modernize language for cetacean science and conservation.

205 The director of your environmental organization is a businessperson, which proves you are only in it for the money.

- How many ministers in your country's government are political appointees rather than true experts in their field?
- Money talks, nobody walks.
- Yes, it may be true that our boss is a finance person, but what do people expect from their top conservation boss? Is it intricate knowledge about a specific species? No. That's rarely a key part of the job description. Scientific excellence alone is insufficient to secure the required scale of resources, partnerships, and policy influence. Accordingly, nature conservation or species protection organizations, like many others, seek leadership experience, strong strategic thinking, exceptional communication skills, and a passion for advancing the

organization's vision and mission. In many cases, this means being the conduit to the outside world (intercultural collaboration) rather than churning out detailed instructions to field workers. Another key skill requirement is keeping the organization afloat financially. And this means that even non-profit organizations need to talk money (which biologists are notoriously poor at). (For a take on funding strategies, see also entry Nr. 175.) Overall, this range of tasks is not necessarily on the radar of your typical hardened, hands-on field crews or specialists.

In an age of enormous competition for funds, along with major budget shortfalls around the world and at all levels of government (not to mention the defunding of science in certain countries), it may help to rethink the relationship between science, conservation, and politics when hiring. The best person to hire and negotiate that minefield may well be someone from outside biology, ecology, or conservation, for example, someone with additional experience in the commercial or policy sectors, someone with business savviness that incorporates conservation needs. And that person has to be paid a competitive top executive salary. (For the reimbursement of the rank and file, see entry Nr. 212.)

Ignacio Jiménez, I., Basurto, X. 2022. An organizational framework for effective conservation organizations. *Biological Conservation* 267: 109471. https://doi.org/10.1016/j.biocon.2022.109471.

20
Activism and Helpful Souls

206. **Don't you get depressed/frustrated seeing how everything is going downhill?**
    - Thanks for that question. It's people like you who care that make it all worthwhile.
    - Yes, but sometimes a glass of beer or a cup of coffee already helps.
    - Let's look at it from a philosophical standpoint. Immanuel Kant raised three core questions for us as members of humanity. These are variously phrased as: "What can I *know*?", "What ought I *do*?", and "What can I *hope* for?". Our conservation work tackles all three. Firstly, we *know* a lot, or let's say we know enough about biology, ecology, and conservation, enough about the animals we seek to protect and about their habitat. And a dollop of common sense also goes a long way. We also know that most people want an intact environment. Secondly, we also know what we have to *do*, and our project is here and now taking the action that needs to be taken (and that has been approved by the responsible authorities). Thirdly, we do *hope* that our efforts and commitment will bear fruit, and the fact that we are here hard at work confirms and strengthens our hope. According to Kant, this perspective underlines the moral duty derived from reason.

https://thomasoppong.com/immanuel-kants-mind-bending-quote-to-shift-your-reality-of-a-good-life/

Kant, I. 1998. *Critique of Pure Reason*. Cambridge University Press. (Original German version 1781). https://philocyclevl.files.wordpress.com›2016/10

DOI: 10.1201/9781003757474-20

207. **What can I do to help your project? (Figure 31).**

Figure 31 The living and working conditions (tents, container units, outdoor offices) for volunteers and other researchers are often spartan at best. Donations or support provided by locals and/or tourists can help make life much more pleasant.

- I've been waiting for years to hear that sentence.
- How about making a donation at our information booth?
- How about inviting my team for dinner in town?
- Do you know a place where we can all take showers?
- Do you have a translator or someone with police connections who could help us resolve that sticky situation we were confronted with last week?
- Here's our bank account number.
- Wow, thanks, yes, we can use more food, equipment, tools, first-aid kits, money, an electricity connection and fuel supply, robust laptops, walkie-talkies, a rain-proof facility, a better boat, a new outboard motor, new diving equipment, an all-terrain vehicle, more powerful weapons (just kidding), new sleeping bags, insect repellent, a stack of local bus tickets, cheap flights, etc., etc.

208. **I know you guys are sea turtle/orang-utan/wolf specialists, but I don't know who else to bring these stray dogs/baby kittens/wounded bird, etc.**
- Yeah, fine, bring your kids, we might be able to treat them too.
- Like most doctors' offices, you can have an appointment in 3 months.
- You mean you touched that bleeding animal with your bare hands!?

- I'm afraid we are only equipped to handle the animals defined and approved in our project and have neither the knowledge nor capacity to treat other animals. And it would actually be illegal for us to do so anyway. Please contact your local veterinarian, wildlife emergency hotline, or other responsible authorities.

209. **Can I bring you something along from town?**
    - Hey, no thanks, but can you give us a ride there and back so that we can do our weekly shopping.
    - Yes, thanks, here's a list (and do I need to give you any money?). Can you come around again next week too?
    - Thanks, we'll be happy to list you in the Acknowledgments section of our future scientific publications.

210. **How can my company help sponsor your project?**
    - That's music to my ears.
    - Thanks for your offer. This would mean so much more than a one-time donation. Species protection/nature conservation projects can only be successful if continued long term. Perhaps we could enter into a longer-term cooperation. Here is my business card. When can our project leaders meet with your team? Can someone in your company help us with a more professional website and fund-raising effort? Can you print T-shirts with our logo on them (never underestimate the power of a free T-shirt!)? Do you have any office space or storage room(s) in town or near our fieldwork site(s)? Do you have any helpful contacts at our fieldwork site(s)?

211. **How can my company benefit if we sponsor your project?**
    - I guarantee you a win-win situation for both of us.
    - Doing the right thing sounds like a good reward.
    - You can mention your support on your website to show how green you are.
    - Let us know what you think we can do for you.
        - Your logo or mission statement on our website?
        - Your logo in all our presentations (and on our T-shirts)?
        - Talks to your employees by our project staff?
        - An opportunity for your employees to visit or participate in our project during their vacation?

212. **I really wanted to volunteer in a conservation project, but I would have earned nothing and had to pay for everything.**
    - Yes, it's called highway robbery.

- Women, children, immigrants – you name it – are regularly exploited around the world, why not budding environmentalists?
- I'm sorry this thwarted your enthusiasm for nature conservation. Unfortunately, many projects run on a shoestring budget, and there just aren't enough funds to adequately pay for every coworker. In the business world, the use of unpaid personnel runs under the moniker of "trainee" or "internship," in environmental matters under "volunteer." Unpaid nature conservation work is irking because most young people wish to (or must) earn some money during the summer. Moreover, the required skills and duties are often very demanding indeed. They tend to conjure up superhuman powers ("leap tall buildings in a single bound . . ."). Here's a selection from typical lists for sea turtle "volunteers":
  - current undergraduate or recent graduate with a biological science-related degree;
  - in good physical condition with the ability to work outdoors under extremely adverse conditions for up to 12 hours (high humidity, rain, heat, and direct sunlight, biting insects);
  - must be very comfortable working alone (independently, with minimal supervision) on a beach at night and around local wildlife;
  - must frequently lift and/or move up to 30 pounds;
  - work all night/into the wee hours of the morning (9 pm–5 am);
  - conduct lengthy daily nesting surveys on foot (or an ATV) starting at sunrise (at work as early as 5 am);
  - preferred proficiency in the language of the country hosting the project;
  - strong communication skills;
  - ability to engage professionally with the public;
  - ability to learn complex tasks quickly and manage a high-stress work environment;
  - attention to detail and strong problem-solving abilities;
  - ability to work independently and in a group;
  - must possess a valid driver's license;
  - must possess a working cell phone;
  - document sea turtle nesting activity and identify each crawl to species and crawl type;
  - tag turtles;
  - locate egg chamber and mark nests for evaluation;
  - conduct nest checks and excavations;
  - document beach obstructions and disorientations;
  - enter and validate data into a GPS unit, on paper data sheets, and a computer database;

- conduct night monitoring work and relocations for sand placement projects;
- assist with stranding reports of sick, injured, or deceased sea turtles;
- conduct sample collection and processing for various research projects;
- contribute and compile content for the Marketing department; and
- be responsible for basic ATV care, including cleaning and fueling.

The only item on the list that's a breeze for most folks these days is a working cell phone. Some countries also require that you be legally authorized to work there. And they all want you to commit for a lengthier period, typically a whole month, but preferably the whole summer or for the full relevant season. Of course, there is no dearth of "volunteer" projects for which the volunteer has to pay handsomely to participate, often in addition to flights and accommodation.

What are the take-home messages? The first is that you have to start your career somewhere, and that may unfortunately involve some sort of volunteer or poorly paid work. Think of yourself as a budding rock musician. Your first gigs are bound to be unpaid (other than, with luck, free beer). The second is that, after doing a stint as a volunteer, you can sell yourself as experienced and advance to a higher, paid rank the next time around. The third is that, if you plan to pursue a degree or a career in nature conservation or related field, you'll be asked what you have done toward that end by the people deciding your fate. How can you convincingly prove your passion? Which candidate is more likely to be accepted or get the position – someone whose summer jobs were being a waiter or salesperson, or someone whose interests are demonstrated by having participated in environmentally relevant activities? Most importantly at any time, the motto is "if you don't ask, you get nothing." Negotiate for free flights/travel expenses. Sometimes these are granted if you sign up for a longer period. Ask for a daily allowance. Ask about free accommodation and meals. Ask about a certification or whether your university will give you credits for participating. Ask whether they will give you a paid position as a field leader or a similar job if you come again next year.

Halpenny, E.A., Caissie, L.T. 2003. Volunteering on nature conservation projects: Volunteer experience, attitudes and values. *Tourism Recreation Research* 28(3): 25–33. https://doi.org/10.1080/02508281.2003.11081414

https://www.thegreenvolunteers.org/

Lithgow, K., Timbrell, H. 2014. How better volunteering can improve conservation: Why we need to stop wondering whether volunteering in conservation is a good thing and just get better at doing it well. *Journal of the Institute of Conservation* 37(1): 3–14. https://doi.org/10.1080/19455224.2013.873730

213. **How can I help save the world/address the extinction crisis/combat climate change/improve my well-being and that of those close to me?**

- Support greater investment in education (especially of girls).
- Lobby all levels of society about nature conservation (starting with your friends and relatives).
- Eat more locally sourced foods.
- Walk, cycle . . .
- Seek regular contact with the outdoors.
- Plant native rather than exotic plants in your garden.
- You're a major landowner? Consider a rewilding project.
- Minimize air travel.
- Reduce home energy use.
- Take the 4–5 "R's" (reduce, refuse, repair, reuse, recycle) and upcycling to heart.
- Invest in institutions that reflect your values.
- Donate to, join, and/or volunteer with groups dedicated to nature conservation.
- Commit to a sponsorship for a zoo or public aquarium animal that you like.
- Who did you say was going to inherit all your money?
- Vote for the right party.
- Read more/stay well informed.
- Invite your discussion partner for a cup of tea/coffee/beer.
- Inspire hope and optimism (see Margaret Mead quote in entry Nr. 201).

Index

For Product Safety Concerns and Information please contact our
EU representative GPSR@taylorandfrancis.com Taylor & Francis
Verlag GmbH, Kaufingerstraße 24, 80331 München, Germany